# Polymer Biodegradation and Polymeric Biomass Valorization

# Polymer Biodegradation and Polymeric Biomass Valorization

Editor

**Piotr Bulak**

MDPI • Basel • Beijing • Wuhan • Barcelona • Belgrade • Manchester • Tokyo • Cluj • Tianjin

*Editor*
Piotr Bulak
Department of Natural
Environment Biogeochemistry
Institute of Agrophysics
Polish Academy of Sciences
Lublin
Poland

*Editorial Office*
MDPI
St. Alban-Anlage 66
4052 Basel, Switzerland

This is a reprint of articles from the Special Issue published online in the open access journal *Polymers* (ISSN 2073-4360) (available at: www.mdpi.com/journal/polymers/special_issues/polym_biodegrad_biomass_valorization).

For citation purposes, cite each article independently as indicated on the article page online and as indicated below:

LastName, A.A.; LastName, B.B.; LastName, C.C. Article Title. *Journal Name* **Year**, *Volume Number*, Page Range.

**ISBN 978-3-0365-7319-9 (Hbk)**
**ISBN 978-3-0365-7318-2 (PDF)**

# Contents

# About the Editor

**Piotr Bulak**

Dr. Piotr Bulak works as an assistant professor in the Department of Natural Environment Biogeochemistry at the Institute of Agrophysics of the Polish Academy of Sciences in Lublin, Poland. His scientific interests include the bioaccumulation of (especially) technologically critical elements in living organisms, in particular in insects; bioremediation, particularly phyto- and entomoremediation; the use of insects for the valorization of waste biomass, including onerous waste; and the biochemistry of abiotic stresses in plants. He is also looking for new ways to use insects or products obtained from them in interdisciplinary areas of science.

# Preface to "Polymer Biodegradation and Polymeric Biomass Valorization"

Both plastics produced from petroleum and naturally occurring polymers such as cellulose or chitin, derived, in turn, from the anabolism of living organisms, are examples of compounds that find a very wide range of everyday use and used in industrial applications. However, the efficient waste management of synthetic polymers has become a huge environmental and economic problem, which best illustrates the ubiquity of microplastics in the environment. The management of waste polymeric biomass appears to be an easy task, but incineration and landfilling are methods that should be limited due to climate change and socio-political pressure. The best example of these trends is the waste hierarchy, determined by the European Union and EPA, in connection to which emphasis should be placed on maximal use of the waste product before it is finally taken out of circulation. The waste should become the raw material.

Biodegradation appears to be a very interesting approach to polymer management, which is well suited to current environmentally friendly expectations. Due to that, the aim of this Special Issue is to present the latest developments in the biodegradation of both synthetic and natural polymers and to provide examples of their valorization in the context of a circular economy.

Example topics include (but are not limited to) the following:

- Current trends and limitations in the biodegradation of waste plastics and biopolymers;

- Management of waste plastics and biopolymers via different treatment technologies and their valorization;

- Development of new bioremediation and biodegradation technologies;

- Recycling, recovery, and valorization of waste polymeric mass;

- Novel and environmentally friendly polymeric materials.

**Piotr Bulak**
*Editor*

 *polymers*

*Article*

# Biodegradation of Different Types of Plastics by *Tenebrio molitor* Insect

Piotr Bulak [1,*], Kinga Proc [1], Anna Pytlak [1], Andrzej Puszka [2], Barbara Gawdzik [2]
and Andrzej Bieganowski [1]

[1] Institute of Agrophysics, Polish Academy of Sciences, Doświadczalna 4, 20-290 Lublin, Poland;
k.proc@ipan.lublin.pl (K.P.); a.pytlak@ipan.lublin.pl (A.P.); a.bieganowski@ipan.lublin.pl (A.B.)
[2] Department of Polymer Chemistry, Institute of Chemical Sciences, Faculty of Chemistry,
Maria Curie-Sklodowska University in Lublin, Gliniana 33, 20-614 Lublin, Poland;
andrzej.puszka@umcs.pl (A.P.); barbara.gawdzik@poczta.umcs.lublin.pl (B.G.)
* Correspondence: p.bulak@ipan.lublin.pl

**Abstract:** Looking for new, sustainable ways to utilize plastics is still a very pertinent topic considering the amount of plastics produced in the world. One of the newest and intriguing possibility is the use of insects in biodegradation of plastics, which can be named entomoremediation. The aim of this work was to demonstrate the ability of the insect *Tenebrio molitor* to biodegrade different, real plastic waste. The types of plastic waste used were: remains of thermal building insulation polystyrene foam (PS), two types of polyurethane (kitchen sponge as PU1 and commercial thermal insulation foam as PU2), and polyethylene foam (PE), which has been used as packaging material. After 58 days, the efficiency of mass reduction for all of the investigated plastics was 46.5%, 41.0%, 53.2%, and 69.7% for PS, PU1, PU2, and PE, respectively (with a dose of 0.0052 g of each plastic per 1 mealworm larvae). Both larvae and imago were active plastic eaters. However, in order to shorten the duration of the experiment and increase the specific consumption rate, the two forms of the insect should not be combined together in one container.

**Keywords:** mealworm; waste management; entomoremediation; bioremediation

**Citation:** Bulak, P.; Proc, K.; Pytlak, A.; Puszka, A.; Gawdzik, B.; Bieganowski, A. Biodegradation of Different Types of Plastics by *Tenebrio molitor* Insect. *Polymers* **2021**, *13*, 3508. https://doi.org/10.3390/polym13203508

Academic Editors: Antonio Pizzi and Beom Soo Kim

Received: 7 July 2021
Accepted: 8 October 2021
Published: 13 October 2021

**Publisher's Note:** MDPI stays neutral with regard to jurisdictional claims in published maps and institutional affiliations.

## 1. Introduction

According to available data from 2019, worldwide plastic production reached 368 Mt (mega tons). The largest amount of plastic was produced in Asia—51% (of which China accounted for 31%), followed by NAFTA countries (North America, Canada, Mexico) 19%, Europe 16%, Middle East, and Africa 7%, Latin America 4%, and Commonwealth of Independent States (CIS) countries 3% [1]. In Europe, of the 29.1 Mt of plastic waste collected in 2018, 32.5% was recycled, 42.6% went to energy recovery, and 24.9% to landfills [1]. Although, as compared to 2016, in 2018 the annual average for recycling and energy recovery of plastic waste increased in Europe by 5.7% and 4.8%, respectively, a significant proportion (24.9%) still end its life cycle in landfills [1]. Furthermore, much of the plastic escapes waste collection systems, polluting the environment and entering the world's oceans. It was estimated that, e.g., in 2010, that of the 275 Mt of plastic produced, as much as 4.8 to 12.7 Mt ended up in the aquatic environment [2]. Adequate waste management policies, as well as the improvement of the plastic waste management methods used, have a huge impact on the environment. In order to reduce the onerous problem of plastic waste, the European Commission has presented a plan which includes, among other things, allowing only reusable or recyclable plastic on the market and recycling at least half the amount of plastic produced [3].

Currently, various forms of recycling are used. This includes mechanical recycling, which involves re-introducing plastic waste into the production cycle and plasticizing it. All thermoplastics such as: polyethylene (PE), polypropylene (PP), and polystyrene (PS)

are suitable for this procedure due to their ability to change shape under high temperature conditions. The disadvantage of this method is that plastics must be carefully sorted before recycling. However, it should be remembered that the food industry uses packaging of a high standard, so in most cases, recovered plastics must not be allowed to come into contact with food. Recycled materials usually go into the production of items of lower value (downcycling), such as garbage bags and traffic cones [4].

Chemical recycling includes activities connected with converting plastic waste into valuable chemical substances or into precursors of plastic. One of the ways can be by pyrolysis, where in anaerobic conditions under the influence of high temperatures and a catalyst plastics decompose, and which contributes to the production of biofuel [5]. Another chemical method is the gasification of plastics, the conversion of waste into gaseous products which can be used in the energy industry (it is necessary to control the level of chlorine content during this process) [6].

Another alternative way to reduce residual waste is to produce plastics capable of degradation under certain natural conditions, such as exposure to sunlight, moisture, having shorter composting times or types of plastics that can be degraded by living organisms. These plastics are called bioplastics, and include polyhydroxyalkanoates (PHA) and polylactide (PLA). Plastics made of petroleum-based polymers, such as PE, PP, PS are difficult to biodegrade. It was long thought that they are not available as carbon source for microbiota. Fortunately, it was found that some microorganisms possess the ability to produce enzymes that break down all these types of plastics [7,8]. The course of biodegradation is influenced by factors, such as the characteristics of the polymer itself (type of chemical bonds) and the storage conditions of the plastic waste. Biodegradation is highly anticipated and future-oriented natural method of plastic waste disposal [9].

There are many studies which show that some bacteria and fungi, especially those associated with the soil, are promising organisms for the biodegradation of plastic waste. The numerous examples include, e.g., the bacteria *Brevibacillus borstelensis* [10], *Bacillus brevis* [11], *Pseudomonas stutzeri* [12], and the fungi *Rhizopus delemar* [13], *Mucor* sp., *Paecilomyces* sp., and *Thermomyces* sp., [14], although more specific research is still needed on all of these, especially in order to increase their biodegradation potential.

Biodegradation potential has also been seen in more complex organisms, such as insects. *T. molitor* has the potential to consume PE, PS, PP, polyurethane (PU), polylactic acid (PLA), and polyvinyl chloride (PVC) [15–21] and even tire crumbs and vulcanized butadiene-styrene elastomer (SBR) rubber [22]. The biodegradability of PE and PS is provided by the larvae of the beetle *Tribolium confusum* [23] and the superworm *Zophobas atratus* [21]. Another example is *Galleria mellonella*, a wax moth, which is capable of PE degradation [24]. Studies have shown that the ability of the insect to biodegrade plastics is in most cases dependent on its intestinal microorganisms. For example, the fungus *Aspergillus flavus* contributed to PE degradation in the abovementioned wax moth [25]. However, studies on *Corcyra cephalonica* rice moth larvae have proven that they have the ability to degrade low density PE (LDPE) even after antibiotic therapy, indicating that biodegradation is not dependent on the activity of intestinal microflora enzymes [26]. These are examples of entomoremediation [27,28] of plastics, a new subtype of bioremediation.

The recent discovery of the land snail *Achatina fulica*, able to disintegrate PS [29] regardless of the composition of the intestinal microflora, in our opinion is an outstanding example of how little we know about the enormous potential hidden in nature when it comes to the utilization of plastics.

*T. molitor* is a species of beetles from the Tenebrionidae family, found in temperate climate regions and considered a storage pest. *T. molitor* larvae is yellow, which is why it is commonly called a "yellow mealworm". Its size varies between 2.5 and 3.5 cm. The adult form is black and measures approximately 1.5 cm [30]. Mealworm larvae are full of nutritional value, making them suitable as live animal feed, e.g., for the zoo industry [31,32]. They are also considered an alternative source of protein in the food industry [33,34]. *T. molitor* frass can be used as a substitute or additive for NPK fertilizers [35], but it can also be used to produce biogas through anaerobic digestion [36]. *T. molitor* itself is likely to have probiotic properties. Studies have shown that adding them to the diet of mice results in the growth of bacteria from the *Bifidobacteriaceae* and *Lactobacillaceae* families in their intestines [37]. Another study has found the enrichment of the beneficial intestinal microflora in rabbits, as well as a reduction in pathogens in their digestive tract [38]. *T. molitor* can also be a source of chitosan production, which can be used to produce UV-protective packaging due to its naturally opaque, brownish color [39].

The aim of the study was, therefore, test the performance of biodegradation of real plastic wastes by *T. molitor*, including polystyrene (PS), two types of polyurethane (PU) and polyethylene (PE) and, for the first time, provided data about the efficiency of polyurethane biodegradation by mealworms. To demonstrate how the process of plastics biodegradation by *T. molitor* really looks like we created time-lapse video. We measured also growth parameters of the insects in order to show how they performed on the wastes and discussed advantages and disadvantages of used approach. Heavy metals and other elements in plastics were also analyzed to see how their content affected the disposal of plastics by mealworms.

## 2. Materials and Methods

### 2.1. Insect Rearing

*T. molitor* larvae were ordered from an external supplier. The larvae were fed with wheat bran (KUPIEC, Poland). The dry weight (DW) of the bran was $87.92 \pm 0.01\%$ ($105\,°C/24$ h) and $pH_{H2O} = 6.56 \pm 0.04$ ($w/v$ 1:20). The mealworms were reared in the following conditions: temperature $24 \pm 1\,°C$, relative humidity $60 \pm 5\%$, in plastic containers at the Department of Natural Environment Biogeochemistry, Institute of Agrophysics PAS, Lublin, Poland. After the acclimatization of the larvae (3 d) they were sifted from wheat brans and left for 24 h to empty their intestines.

### 2.2. Plastic Waste

Four types of plastics were tested in the experiment:

1.  Polystyrene (PS) in the form of Styrofoam, which is used for insulating building elevations;
2.  Polyurethane foam (PU1)—in the form of kitchen sponges;
3.  Polyurethane thermal insulation foam (PU2)—consisting of polyphenylpolyisocyanate polymethylene, according to information from the producer (SOUDAL Sp. z.o.o, Poland);
4.  Polyethylene foam (PE)—which is used as filling material in packages, e.g., to protect electrical equipment during transport.

The plastic used is showed on Figure 1. Kitchen sponge (PU1) was used without the abrasive layer, which is often added to that type of houseware item. A day before the beginning of the experiment, PU2 was removed from the original packaging and left to bind in the form of one large mass. The following day, a thick slice was cut from the inside of this piece and was used in the experiment. All plastics were cut to obtain an equal weight of 2.6 g ($\pm 0.001$ g) before the experiment begun, i.e., to match to the weight of the PU1 (kitchen sponge), which was taken as a whole

The original plan of the experiment assumed the use of only PS and two types of PU, but during the course of the experiment, we decided to add one more variant with PE foam, as there was no research on the possibility of PE foam biodegradation by mealworm larvae at that time. However, adding the PE variant was not possible in the experiment, which was just being filmed, so the time-lapse video shows only 3 types of plastic.

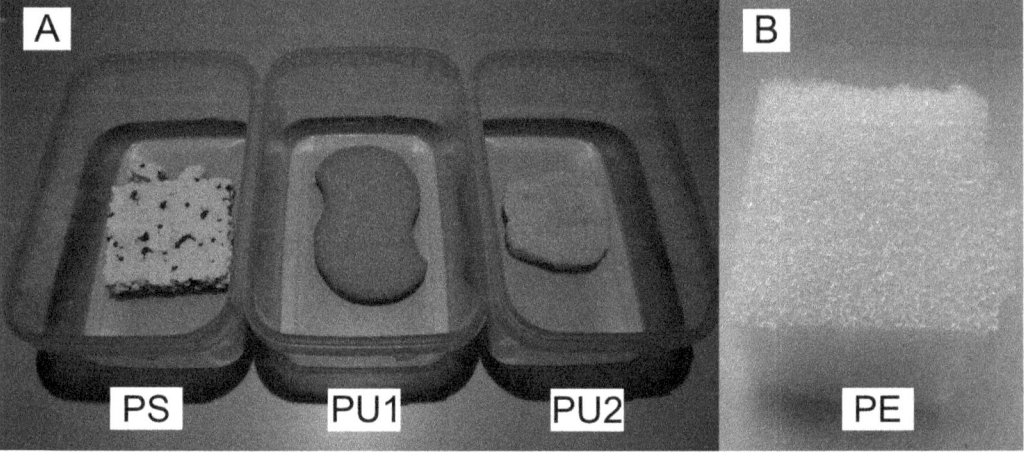

**Figure 1.** Plastics used for the experiment. (**A**) Materials for which time-lapse video was created. From left: polystyrene (PS), polyurethane kitchen sponge (PU1), polyurethane thermal insulation foam (PU2). (**B**) Polyethylene packaging foam (PE).

*2.3. Experimental Procedure*

Each of the four experimental variants was carried out in three independent replications in plastic boxes with dimensions of $30 \times 19 \times 12$ cm (see Figure 1). The experiment was conducted in a laboratory room under controlled conditions: temperature $24 \pm 1$ °C and relative humidity $60 \pm 5\%$. Each container contained 500 mealworm larvae with a length and weight of individual larvae about $2.39 \pm 0.02$ cm and $0.100 \pm 0.010$ g, respectively. All variants (PS, PU1, PU2, PE) included 2.6 g of a given plastic, therefore at the beginning of the experiment 0.0052 g of each plastic was available per 1 mealworm larvae (based on literature data and our preliminary trials). The substrates were not enriched with additional substances such as water or feed.

The substrates were weighed daily on a laboratory balance (OHAUS EX224M, Parsippany, NJ, USA) in order to investigate the loss of plastic. During the experiment, when the pupae started to appear massively, they were taken out of the container with tweezers and put into a separate glass vessel for pupation. After completion of this process, the adult beetles were once again moved back into the original container. This operation can be seen on the recorded Video S1. Adult beetles are also capable of eating plastics and can survive on this substrate as the sole source of food, similar to the larvae. As mealworm pupae are not mobile, removing the pupae during the experiment was performed in order to protect them from cannibalism by other developmental stages of the insect. Cannibalism, which is always present in mealworm breeding, would reduce the amount of eaten plastic.

The experiment lasted 58 days and was completed when the consumption of plastics by the insects became minimal. After that time the weight of the insects was determined by the use of a laboratory balance and their length using a hand ruler. In order to separate insect excrement from the uneaten plastic remains, they were sieved through a 500 μm mesh size sieve.

The utilization rate (U) of the plastics was calculated based on the following Equation (1):

$$U = \frac{m_i - m_f}{m_i} \cdot 100\% \tag{1}$$

where:
$m_i$—initial mass of plastic (i.e., at the beginning of the experiment) (g);
$m_f$—final mass of plastic (i.e., at the end of the experiment) (g).

Plastic remnants that passed through insect digestive tract were extracted from the mealworm frass collected after the experiment. One gram of the frass from each experimental variants was demineralized with 5% HCl by 1 h in room temperature and then digested using 30% $H_2O_2$ by 48 h in room temperature. Digestates were filtered on glass filter with pore size 1 μm [40]. After that the samples were dried in 50 °C for two days (until no mass changes were noticed).

### 2.4. Fourier Transform Infrared Spectroscopy (FT-IR) of Plastics

The FT-IR spectra were developed by applying attenuated total (internal) reflection (ATR/FT-IR) with the use of a FT-IR TENSOR 27 spectrophotometer (Bruker, Germany), complete with a *PIKE* measuring cell which features crystalline diamond embedded in zinc selenide. The FT-IR spectra were collected within the range of 4000 to 600 cm$^{-1}$, with 32 scans per sample, at a resolution of 4 cm$^{-1}$. The absorption mode was used for these measurements. Figure 2 showed spectra obtained for used materials.

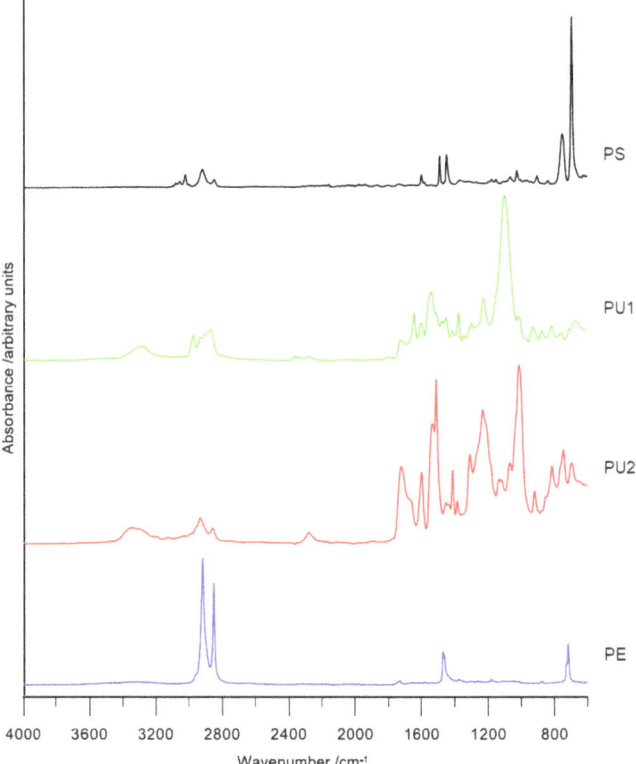

**Figure 2.** FT-IR spectra for plastics used in the experiments. PS—polystyrene (Styrofoam), PU1—polyurethane foam (kitchen sponge), PU2—polyurethane foam (building thermal insulator), PE—polyethylene foam (packaging foam).

The FT-IR spectra of the tested plastics are shown in Figure 2. The peaks at 2921 cm$^{-1}$ and 2850 cm$^{-1}$ correspond to C–H asymmetric and symmetric stretching in $CH_2$ groups. The presence of a benzene ring in PS is confirmed by absorption bands at 3060 cm$^{-1}$ and 3026 cm$^{-1}$ (correspond to C–H stretching in benzene ring), band at 754 cm$^{-1}$ corresponds to out-of-plane C–H bending of the benzene ring. Absorption peaks at 1601 cm$^{-1}$, 1493 cm$^{-1}$,

1452 cm$^{-1}$ are related to C=C stretching of the benzene ring. The absorption band at 696 cm$^{-1}$ with high intensity corresponds to a monosubstituted benzene ring.

The FT-IR spectra of both polyurethane materials show characteristic absorption bands responsible for N–H stretching vibrations (in the range 3290 to 3340 cm$^{-1}$), N–H bending (at 1509 cm$^{-1}$) and C=O stretching (in the range 1727 to 1721 cm$^{-1}$) of the urethane group and absorption bands at 2929 cm$^{-1}$ and 2868 to 2855 cm$^{-1}$ (asymmetrical and symmetrical, respectively) corresponding to C–H stretching vibrations of the CH$_2$ group, and at 1383–1373 cm$^{-1}$ corresponding to C–C bending vibrations.

The spectra of PE show characteristic absorption bands at around 2915 cm$^{-1}$ and 2849 cm$^{-1}$ connected to C–H stretching in the CH$_2$ group while absorption band at 1471 cm$^{-1}$ correspond to C–H bending in CH$_2$ group. Absorption bands at 729 cm$^{-1}$ and 718 cm$^{-1}$ are connected to C–H bending in-plane in CH$_2$ groups.

### 2.5. Energy Dispersive X-ray Fluorescence (EDXRF)

The concentrations of elements in the plastics samples were measured by means of EDXRF using EDX-7000 (Shimadzu, Kioto, Japan). Measurements were completed in air atmosphere using default plastics measuring program with autobalance. Measurement time was 100 s on each channel and collimator had 10 mm ϕ. Samples were placed directly in the Rh X-ray beam without the use of foil or special containers.

### 2.6. Scanning Electron Microscopy (SEM)

Plastics samples taken for the experiment, as well as plastics remnants isolated from the insect frass after the experiment were sputter coated with a 30 nm Au layer in the coater EM ACE (Leica, Germany) and visualized by the use of Libra SEM (Carl Zeiss, Germany) set to the following parameters: work distance (WD) 9–15 mm; spot size 360–370, accelerating voltage 5 keV; aperture size 30 μm; beam current 30 μA; signal SE detector with scanning mode and pixel noise reduction.

### 2.7. Photography and Time-Lapse Movie of Utilization

Photography was completed with the use of a Nikon D7100 with an AF-D DX NIKKOR 18-105mm f/3.5-5.6G ED VR lens. Photographs were taken automatically every 30 min and a time-lapse movie was created from them using Adobe After Effects 2020 and Adobe Media Encoder 2020.

### 2.8. Statistical Analysis

Experimental results were analyzed using Statistica 13.1. The statistical significance was determined by t-Student test and ANOVA with post hoc Tukey's test ($p < 0.05$; n = 3). Three independent biological replications of the experiment were performed.

### 3. Results

#### 3.1. Rate of PS, PU1, PU2, and PE Consumption by Mealworms

Figure 3 illustrates the weight reduction in plastics by *T. molitor* insect. The first significant (Student *t*-test, $p < 0.05$) loss of mass for each type of plastic was observed on the third day of the experiment and amounted to 0.566, 0.569, 0.482, and 0.313 g (PS, PU1, PU2, and PE, respectively). On the 15th day, the most uniform utilization values of individual plastics were observed, ranging from 0.85 to 0.94 g. On the 58th day of the experiment, statistically significant differences in the reduction mass of the plastics were shown and were as follows: PS 1.385 g, PU1 1.221 g, PU2 1.538 g, and PE 1.818 g, which is equivalent to a percentage loss of 46.93 ± 0.12%, 46.77 ± 2.73%, 58.97 ± 5.15%, and 69.71 ± 6.34%, respectively (Table 1) (PE > PU2 > PS > PU1). Interestingly, at the beginning, the utilization of PE was the lowest, however, it was ultimately the highest of all the plastics used in this trial.

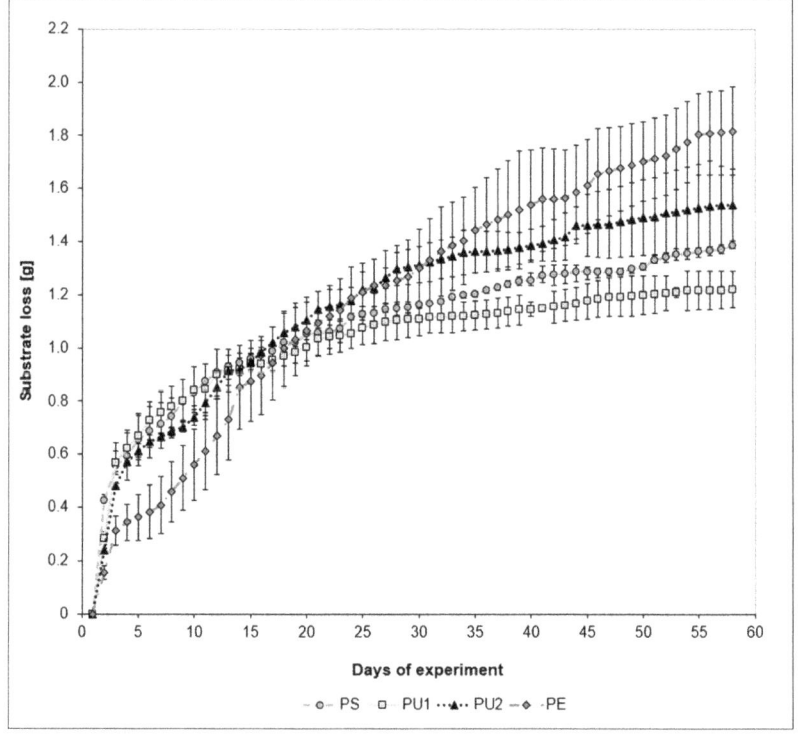

**Figure 3.** Cumulative loss of PS, PU1, PU2, and PE plastic waste (g) during feeding of *T. molitor* (mean ± SD; n = 3).

**Table 1.** Weight of the plastics at the beginning and end of the experiment and the mass of plastics remnants in the mealworm frass (mean ± SD; n = 3).

| Parameter\Plastic Type | PS | PU1 | PU2 | PE |
|---|---|---|---|---|
| Initial | | | | |
| **Mass of plastic [g]** | 2.610 ± 0.001 | 2.611 ± 0.001 | 2.607 ± 0.001 | 2.608 ± 0.001 |
| Final | | | | |
| **Mass of plastic [g]** | 1.225 ± 0.004 [b,c,*] | 1.390 ± 0.072 [c,*] | 1.070 ± 0.133 [a,b,*] | 0.790 ± 0.166 [a,*] |
| **Utilization [*w/w* %]** | 46.929 ± 0.124 [a,b] | 46.771 ± 2.734 [a] | 58.972 ± 5.146 [c] | 69.707 ± 6.341 [c] |
| **Plastic remnants in frass [*w/w* %]** | 31.697 ± 0.647 [a] | 45.353 ± 1.548 [d] | 41.724 ± 0.543 [c] | 38.977 ± 1.435 [b] |

Significant differences between initial and final values of a given parameter are indicated with * (Student *t*-test, $p < 0.05$). Post hoc Tukey's test was completed to show significant differences between all the plastics within a given variable (different letters; $p < 0.05$).

### 3.2. Morphological Parameters of Mealworm Larvae

Table 2 shows morphological parameters of *T. molitor* larvae. At the end of the experiment, the mass of 1 larvae in all variants, i.e., PS, PU1, PU2, PE decreased by 18.37%, 28.28%, 26.26%, and 24.71%, respectively. Similarly, decrease in larvae length was also observed and amounted for PS 6.26%, for PU1 9.35%, for PU2 4.35%, and for PE 6.58% (Table 1).

**Table 2.** Average length and fresh weight of the insects at the beginning and end of the experiment ($\pm$SD; n = 3).

| Parameter\Substrate | PS | PU1 | PU2 | PE |
|---|---|---|---|---|
| | | Initial | | |
| Mass of 1 larvae [g] | 0.098 ± 0.001 | 0.099 ± 0.001 | 0.099 ± 0.001 | 0.085 ± 0.002 |
| Length of 1 larvae [cm] | 2.413 ± 0.013 | 2.397 ± 0.060 | 2.367 ± 0.003 | 2.386 ± 0.047 |
| | | Final | | |
| Mass of 1 larvae [g] | 0.080 ± 0.002 [b,*] | 0.071 ± 0.003 [a,b,*] | 0.073 ± 0.006 [a,b,*] | 0.064 ± 0.001 [a,*] |
| Length of 1 larvae [cm] | 2.262 ± 0.005 [a,*] | 2.173 ± 0.010 [a,*] | 2.264 ± 0.093 [a] | 2.229 ± 0.005 [a,*] |
| Mass of 1 adult [g] | 0.120 ± 0.001 [b] | 0.114 ± 0.008 [b] | 0.089 ± 0.001 [a] | 0.087 ± 0.007 [a] |
| Length of 1 adult [cm] | 1.617 ± 0.017 [b] | 1.450 ± 0.050 [a] | 1.459 ± 0.092 [a] | 1.322 ± 0.006[a] |

Significant differences between initial and final values of a given parameter are indicated with * (Student *t*-test, $p < 0.05$). Post hoc Tukey's test was completed to show significant differences between all the plastics within a given variable (different letters; $p < 0.05$).

### 3.3. Elements Content in the Plastics

Investigated plastics had in general trace amounts of different elements (Table 3), with some exceptions. PS had the highest concentration of Br, amounted to 0.412 ± 0.019%. The samples of polyurethane kitchen sponge (PU1) was characterized by high concentration of Ca (10.110 ± 0.301%), Al (0.437 ± 0.038%), Si (0.358 ± 0.027%), and Cl (0.224 ± 0.022%). The second investigated polyurethane material (PU2) had different elemental composition and characterized by high content of Cl (7.574 ± 0.141%) and P (1.379 ± 0.035%). PE had only minor content of Al (0.180 ± 0.029%) and Ca (1.381 ± 0.020%). Interestingly in PS and PE trace amounts of Hf was detected.

**Table 3.** The concentrations of elements measured in plastics samples by means of EDXRF ($\pm$ SD; n = 3).

| % | PS | PU1 | PU2 | PE |
|---|---|---|---|---|
| Al | - | 0.437 ± 0.038 | - | 0.180 ± 0.029 |
| Ba | - | - | - | 0.029 ± 0.001 |
| Br | 0.412 ± 0.019 | - | 0.001 ± 0.001 | - |
| Ca | 0.020 ± 0.009 | 10.110 ± 0.301 | - | 1.381 ± 0.020 |
| Cl | 0.017 ± 0.001 | 0.224 ± 0.022 | 7.574 ± 0.141 | 0.017 ± 0.004 |
| Co | 0.002 ± 0.001 | - | - | - |
| Cr | 0.003 ± 0.001 | 0.008 ± 0.001 | 0.001 ± 0.000 | 0.004 ± 0.001 |
| Cu | 0.006 ± 0.000 | 0.013 ± 0.001 | 0.001 ± 0.000 | 0.006 ± 0.001 |
| Fe | 0.008 ± 0.002 | 0.021 ± 0.001 | 0.001 ± 0.000 | 0.006 ± 0.001 |
| Hf | 0.010 ± 0.001 | - | - | 0.012 ± 0.000 |
| K | 0.009 ± 0.005 | - | 0.004 ± 0.002 | 0.005 ± 0.001 |
| Mn | 0.003 ± 0.001 | 0.007 ± 0.001 | - | 0.002 ± 0.000 |
| Ni | - | 0.001 ± 0.001 | - | - |
| P | - | - | 1.379 ± 0.035 | - |
| S | 0.023 ± 0.002 | 0.008 ± 0.005 | - | 0.012 ± 0.003 |
| Si | 0.049 ± 0.005 | 0.358 ± 0.027 | 0.072 ± 0.004 | 0.029 ± 0.004 |
| Sn | - | 0.070 ± 0.000 | - | - |
| Ti | - | - | - | 0.004 ± 0.000 |
| Zn | 0.003 ± 0.001 | 0.003 ± 0.000 | - | 0.007 ± 0.000 |

*3.4. Scanning Electron Microphotography of the Plastics Surfaces*

Figure 4 showed SEM microphotography taken from the plastic samples, which were used in the experiment and those which were isolated from the insect frass. At the beginning all materials were characterized by smooth surfaces. PS (Figure 4A) exhibited a smooth and delicately cross-linked surface structure with visible longitudinal cracks arranged in the same direction which were most likely caused by mechanical impact during sample preparation (breaking and cutting of brittle material). PS isolated from the frass showed completely different, rough and irregularly carved surface structure (Figure 4B). PUs foams (Figure 4C,E) structures were at the beginning also smooth but PU1 (Figure 4C) was more flat while PU2 (Figure 4E) showed a greater number of edges. After the passage of the digestive system of insects, the surface of the PU foams was significantly wrinkled (Figure 4D,F) and the edges clearly visible in the PU2 structure at the beginning were significantly smoothed. At the magnification of 20,000× PE surface consisted of the flat valleys and flaky layers of material superimposed on one another with visible fine pores present in a small amount (Figure 4G). After digestion by the insect PE had undergone wrinkles, the pores were no longer visible and instead of them numerous small bubble-like protrusions have appeared (Figure 4H).

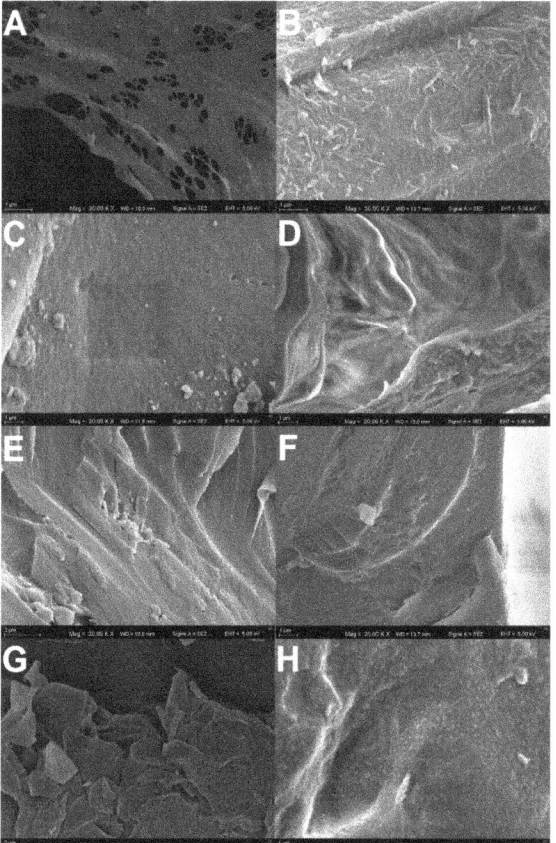

**Figure 4.** SEM microphotography of the plastic samples taken for the experiment (**A,C,E,G**) and plastics isolated from the insect frass after the experiment (**B,D,F,H**). (**A,B**): PS (30,000×); (**C,D**): PU1 (20,000×); (**E,F**): PU2 (20,000×), (**G,H**): PE (20,000×). Magnifications were selected to best illustrate the differences in the surface of the samples.

*3.5. Time-Lapse Movie of Utilization*

Video S1 shows the setup used for this experiment. The initial mass of each plastic (2.6 g) was adopted due to the weight of the PU1, which was a whole unused kitchen sponge. We tried to adjust the mass of the other plastics so that they were in one piece, but it was not always possible, and some required more material to be added (as seen in the example of PS). The sharp changes in the position of the plastic pieces visible in the video resulted from weighing them during the experiment in order to obtain data for Figure 3.

## 4. Discussion

*4.1. Rate of PS, PU1, PU2, and PE Consumption by Mealworms*

The results of PS have shown a higher degree of utilization compared to the studies of [16], where the reduction was only 9.0%, and [20,41], where the reduction was 31.0% and 39.1%, respectively. Mealworms from three Chinese regions of Guangzhou, Tai'an, Shenzhen utilized 57.5%, 34.4%, and 52.4% of PS, respectively [19]. In the study [42], authors used two mealworm species (*T. molitor* and *T. obscurus*) and reached 41.5% and 55.4% of PS utilization. They also found that the application of food additives in the form of wheat bran for *T. molitor* and corn flour for *T. obscurus* resulted in increased PS utilization: 56.8% and 67.1%, respectively. A similar increase was obtained by [43], where the addition of wheat bran to PS increased the utilization rate from 31.7% PS to 54.4%.

To our best knowledge there are no reports about the efficiency of PU utilization by mealworm. The researchers [44] investigated the epigenetic modification of mitochondrial DNA in *T. molitor* caused by PU as a sole source of feed, however they did not present any utilization parameters. Others [45], applied another insect species, *Zophobas morio*, from the same family as *T. molitor* (Tenebrionidae) to utilize PU, but the degradation was only 6%. Our results showed a remarkably higher efficiency. Utilization of PU1 (kitchen sponge) and PU2 (commercial insulation foam) (Figure 3, Table 1) amounted to 46.8% and 59.0%, respectively. This significant difference between both materials was probably caused by the different chemical composition (Figure 2) and, hence, its different macrostructure and properties (e.g., hardness). PU1 was soft and elastic, while PU2 was rigid and brittle. The observations show that regardless of the material itself, it is much easier for mealworms to bite and chew something that is stiff and brittle than flexible and soft. FTIR spectra showed the presence of different additives in both PUs (Figure 2), while element content analysis (Table 3) allow to deduce to which classes of chemical compounds these additives can belong (discussion below, see Section 4.3). Moreover, FTIR spectra confirmed that PU1 was soft PU due to the presence of the 1120 cm$^{-1}$ band characteristic for aliphatic alcoxy-groups (Figure 2). PU2 was rigid foam due to presence of the band around 1600, 100, and 650 cm$^{-1}$, which confirmed ring structure of PU2 formulation (Figure 2). The spectrum showed also 2 characteristic bands at 1250 and 1040 cm$^{-1}$, which indicated the aromatic alkoxy-groups (Figure 2). Soft foams are formed of polyols with a molecular weight of 2000 to 8000 units and diisocyanates, while rigid foams are made of polyols with a molecular weight of less than 1000 units and a mixture of di- and triisocyanates. Summarizing, the differences in the utilization of PU1 and PU2 resulted from the different chemical composition of both materials, as well as the difference in mechanical properties. The results obtained for PE reduction were higher than the results presented in the literature. [19] exploited the mealworm larvae for PE disposal and showed that the origin of the larvae (different regions in China) influenced the degree of utilization. The results were: 36.9% loss of PE for the larvae from Guangzhou, 22.0% for larvae taken from Tai'an, and 29.7% for larvae from Shenzhen. These differences were probably caused by the different microbiomes inhabiting the intestines of the larvae. The addition of wheat bran increased the degradation rate of PE from 48.3% to 61.1% [43].

It must be mentioned that not 100% of the eaten amount of given plastic is assimilated by the mealworms. As revealed by [20], 47.7% of ingested PS carbon was converted to $CO_2$ and ca. 0.5% was assimilated into biomass (lipids). This indicated that ca. 50% of PS left in the frass [20]. Our results showed that the total amount of the plastics remnants in the mealworm frass was in the range of 31.7% to 45.4% (Table 1). PS was the most susceptible to biodegradation while PU1 the least. The relatively high presence of plastics in the feces is disadvantageous due to the possibility of spreading microplastics, however research is needed on the susceptibility of these residues to microbial degradation, as it can be changed due to the action of enzymes in the insects digestive tract.

Table 4 showed specific consumption rate given as µg plastic·day$^{-1}$·larvae$^{-1}$ calculated for our results, as well as for literature data. In most of the articles, the experiments were conducted for 30 days, therefore to calculate this parameter we used our data for plastic waste utilization also for that day. In general, specific consumption rate for PS as reported in the literature was within the range of 118.0 to 268.2 µg plastic·day$^{-1}$·larvae$^{-1}$ (Table 4) and can be enhanced by the addition of more natural and nutritious feed, such as soy protein, up to 491.0 µg plastic·day$^{-1}$·larvae$^{-1}$ [46]. PE consumption rates were in the lower range than PS: 102.7 to 226.6 µg plastic·day$^{-1}$·larvae$^{-1}$ (Table 4) and was increased to 286.5 µg plastic·day$^{-1}$·larvae$^{-1}$ by the addition of wheat bran [47]. Lower than reported so far consumption rates calculated for our results were the consequence of experiment design, in which both larvae and adult insects were used in one container to eat plastics. This increased the chance of cannibalism occurring. The second reason was the early appearance of pupae (which can be clearly seen in Video S1), which excluded a large number of insects from actively eating plastic during pupation. Pupae started to appear in large numbers starting from day 10. Specific consumptions rate calculated for this day fall within the range reported in the cited literature (Table 4).

**Table 4.** Plastic consumption rates calculated on the basis of literature data and from this publication.

| Plastic | Consumption Rate [µg·day$^{-1}$·larvae$^{-1}$] | Literature |
|---|---|---|
| Polystyrene (PS) | *T. molitor* from Guangzhou: 268.3<br>*T. molitor* from Tai'an: 160.5<br>*T. molitor* from Shenzhen: 239.9 | [19] |
| | 119.9 | [20] |
| | *T. molitor*, PS alone: 243.0<br>*T. molitor*, PS + wheat bran: 332.3<br>*T. obscurus*, PS alone: 324.4<br>*T. obscurus*, PS + corn flour: 392.4 | [40] |
| | PS alone: 118.0–222.0<br>PS + soy protein: 491.0<br>PS + wheat bran: 441.0 | [46] |
| | PS alone: 148.4<br>PS + wheat bran: 255.2 | [47] |
| | 77.4 [a]<br>174.7 [b] | This research [a] |
| Polyurethane (PU) | PU1: 73.9 [a]<br>168.7 [b]<br>PU2: 87.5 [a]<br>158.9 [b] | This research [a] |
| Polyethylene (PE) | *T. molitor* from Guangzhou: 172.2<br>*T. molitor* from Tai'an: 102.7<br>*T. molitor* from Shenzhen: 138.6 | [19] |
| | PE alone: 226.6<br>PE + wheat bran: 286.5 | [47] |
| | 86.7 [a]<br>122.4 [b] | This research [a] |

a—calculated from the weight loss data of the given plastic on the 30th day of the experiment (Figure 3).
b—calculated from the weight loss data of the given plastic on the 10th day of the experiment (Figure 3).

Genetic differences between mealworm populations around the globe can be the third reason. As shown in [19], genetic differences can be significant even between national populations.

From our results it can be estimated how many mealworm insects (in terms of pieces and mass) would be needed to utilize 1 kg of each waste plastic in the same time as in the presented experiment (i.e., 58 days). During the experiment, 500 insects consumed 1.385 g of PS, 1.221 g of PU1, 1.537 g of PU2, and 1.818 g of PE. Approximating directly, the following number of insects would be needed: 361,011 pc. for PS, 409,500 pc. for PU1, 325,309 pc. for PU2, and 275,028 pc. for PE. The average mass of 500 pc. of larvae we used for the experiment was 56.166 g. Therefore, in terms of mass, one would need the following number of mealworms for the utilization of 1 kg of PS, PU1, PU2, PE during 58 days: 40.5 kg, 46.0 kg, 36.5 kg, and 30.9 kg, respectively. Such amounts are currently commercially available from large growers.

### 4.2. Morphological Parameters of Mealworm Larvae

Different studies reported different changes in mass. A decrease in the average weight of *T. molitor* larvae fed with PS differed greatly between the larvae from different sources and amounted to 3.31%, 21.70%, and 37.06% for Guangzhou, Tai'an, and Shenzhen, respectively [19]. Similar results were obtained by [42], where the reduction in *T. molitor* larvae weight was 8.6% on PS. However, in the study of [41], no significant changes in weight of larvae fed with PS were observed. Contrary, a weight increase of 2.5 ± 1.0% was observed by [46] for expanded PS.

Decrease in the mass of mealworm larvae fed with PE, which amounted to 1.03%, 22.10%, and 24.87% was also noticed for the abovementioned Chinese regions, respectively [19]. Ref. [46] used two types of PE: PE1, which had added pink colorant and PE2, which was without colored additives. On the first material, an 8.8 ± 2.1% increase in larval weight was observed and on the second one a decrease of 3.4 ± 1.6%. This allows to conclude that differences may at least be partially dependent on additives for plastics, such as colorants and fillers.

Unfortunately, the cited research did not present the changes in insect length. Literature data [20] suggested that, most likely, the insects lose some of the fat tissue they had before the experiment and the hydration level of their bodies decreases significantly. This is because, in the studies, plastics were usually the only source of food and larvae had no access to water. Recently, ref. [47] showed that mealworm fed with polystyrene did indeed have a lower fat content than those who ate a conventional diet. Amazingly, they were still capable of successful pupation, which indirectly proves that digestion and assimilation must occurred.

### 4.3. Elements in the Plastics

To check whether differences in the content of elements may affect the utilization of plastics by mealworm larvae, their content was examined. Not even trace amounts of toxic heavy metals, such as Cd, Hg, or Pb, were identified in any of the samples. The majority of elements were present only in trace amounts (much below 0.1%).

Among all 13 identified elements in PS samples, only Br had non-trace concentration of 0.4%. The compound containing Br, which are added to plastics fall into class of flame retardants (FR) and green and red pigments [48,49]. The presence of Br can be connected with the addition of, e.g., hexabromocyclododecane, which is widely utilized FR in the formulations of PS [48]. Kitchen sponge (PU1) had the highest concentration of Ca. Ca in plastics serves, mainly, as a filling and reinforcement agent, which is added in the form of $CaCO_3$ [48]. PU1 had also ca. 0.6% Si, which indicated that Ca may have been partially added as wollastonite ($CaSiO_3$) or other Si compounds, which improved FR properties of the plastic [50]. The content of Cl in PU1 may be connected with the addition of green Cl-based pigments, chemically belonging to chlorinated/brominated phthalocyanine compounds [48]. Additionally, Cl-containing compounds, such as, e.g., Triclosan, could be added to PUs as antimicrobial agents [48]. Much higher content of

Cl found in PU2 altogether with 1.4% P may be the result of using as FRs chlorinated alkyl phosphates [49], such as, e.g., tris(1-chloro-2-propyl) phosphate and tris-2-chloroethyl phosphate; such compounds are used in rigid and flexible PU foam formulations [48]. PU1 had also elevated concentration of Al. The compounds of these element are used, i.e., as stabilizers and FR but in much higher concentrations, therefore the presence of Al in PU1 was probably due to the use of Al-based fillers [48].

The presence of Ca in PE was the most likely connected with the inorganic filler added to the plastic to enhance its strength and decrease the cost of production, while low amount of Al may result from the use of compounds, such as mica or kaolin to surface stabilization of the material [48].

Analysis showed trace amount of quite exotic Hf metal in PS and PE samples. In PE it can be connected with the use of catalyst based on Hf metallocene compounds, which are widely used in the production of polyolefins [51]. In PS, it can be the results of contamination during production of this plastic by a manufacturer, which also produced polyolefins.

Figure S1 showed concentration of elements in the larvae after the experiment determined by EDXRF. Regardless of the variant, major mineral components of the larvae were K, Cl, P (>1%) and S (1–0.5%). Ca was present in concentrations of 0.1%, Zn and Fe in the order of 0.02–0.01% and Cu below 0.01%. Non-physiological elements present in the plastics probably not affected the mealworms due to its low concentrations.

### 4.4. Polymers Surface Alterations Visualized by SEM

Surface alterations of the plastics are characteristic of the aging processes occurring under the influence of microorganisms (present in the intestines of insects) or, more specifically, enzymes secreted by them. The folding of the previously smooth surface of the polymer and the formation of pitting was observed during the biodegradation of PS and PE with the bacterium *P. aeruginosa* isolated from the gut of *Zophobas morio* insect (Tenebrionidae; cousin of mealworm, known also as superworm) [52]. Surface elements mapping comparison of PS and PE before and after biodegradation showed that those changes were connected with increase in oxygen content, which suggests oxidative changes in the plastics [52]. Similar changes as observed by us in the case of PU foams had been observed by [53] after biodegradation trials in soil burial. SEM surface microphotography (Figure 4) was indirect proof that plastics eaten by the mealworm biodegraded during the action of microorganisms enzymes, as well as enzymes secreted by mealworms itself.

### 4.5. Time-Lapse Movie of Biodegradation

It can be seen that during the experiment, pupae started to appear quite quickly, especially on PU1 (Video S1). It was about 10 days after experiment had started. At the beginning, we had not planned to pull out the pupae but it became obvious that without this step the degradation of the plastics would be lowered due to the cannibalistic behavior of the larvae. This is a known problem in mealworm breeding, especially when the insects have limited access to water and a low protein content in the substrate [54]. One of the aims of this study was to reach as a high a utilization of plastics as possible, therefore we decided to take advantage of the adults' ability to feed on plastics as well. When the amount of the adult insects increased in the containers, cannibalism started to occur with the larvae being the victim. It can be seen in the second part of the Video S1, adults formed specific "bundles" for a while with eaten larvae inside it. During the experiment dead adults were taken out the of the containers to prevent the larvae from preferring them as food. Our observations suggests also that the lifespan of the adults was much shorter than it would have been on optimal feed. Video S1 suggested that the adults' ability to consume plastic can also be useful but they should not be combined with earlier stages within the one container to prevent cannibalism and decrease the utilization of plastics.

## 5. Conclusions

The efficiency of mass reduction for all of investigated plastics was 46.5%, 41.0%, 53.2%, and 69.7% for PS, PU1, PU2, and PE, respectively. However, specific consumption rates for each plastics was lower that calculated from literature data. This was due to the large number of pupae appearing less than two weeks after the start of the experiment and the combination of larvae with adults in one container, which resulted in cannibalistic behavior. Additionally, the used plastic waste were characterized by the addition of fillers and FRs, which may influenced consumption rates. The utilization of plastics can be increased by removing pupae from larvae and imago and by not combining adult and larval forms in one container. Both larvae and imago were active in the eating of plastics. More research is needed on different optimization approaches, which would reduce the number of insects used while maintaining process efficiency. Such optimization should be completed in order to decrease the costs of entomoremediation for larger amounts of plastics. The risk of spreading microplastics with insect feces which left after this process should also be determined in the future research.

**Supplementary Materials:** The following are available online at https://www.mdpi.com/article/10.3390/polym13203508/s1, Figure S1: Element contents in *T. molitor* larvae after the experiment, Video S1: Time-lapse movie of the process of plastics waste entomoremediation by *Tenebrio molitor*.

**Author Contributions:** P.B.: Conceptualization, Methodology, Formal analysis, Investigation, Writing—original draft, Writing—review and editing, Supervision, Funding acquisition, Project administration. K.P.: Investigation, Methodology, Writing—original draft. A.P. (Anna Pytlak): Methodology, Writing—original draft. A.P. (Andrzej Puszka): Methodology, Investigation, Writing—original draft. B.G.: Methodology, Supervision. A.B.: Methodology, Supervision. All authors have read and agreed to the published version of the manuscript.

**Funding:** This research was partially founded by the National Science Centre, Poland, grant number 2019/35/D/NZ9/01835. The APC have been granted by MDPI.

**Institutional Review Board Statement:** Ethical review and approval were waived for this study, as the use of invertebrate animals, like insects, for laboratory experiments do not require approval of bioethics committees and are not subject to procedures in accordance with Directive 2010/63/EU of the European Parliament and of the Council of 22 September 2010 on the protection of animals used for scientific purposes, amended by Regulation (EU) 2019/1010 of the European Parliament and of the Council of 5 June 2019 as well as Polish National Act of 15 January 2015 on the protection of animals used for scientific or educational purposes.

**Informed Consent Statement:** Not applicable.

**Data Availability Statement:** Data may be provided upon request to the corresponding author.

**Acknowledgments:** The authors wish to thank Katarzyna Domańska for her help in preparing the time-lapse movie and Tomasz Skrzypek for SEM photography of the samples.

**Conflicts of Interest:** The authors declare that they have no known competing interest. The funders had no role in the design of the study; in the collection, analyses, or interpretation of data; in the writing of the manuscript, or in the decision to publish the results.

## References

1. Plastics-the Facts 2020. An analysis of European Plastics Production, Demand and Waste Data. Plastics Europe, Association of Plastics Manufacturers. 2020. Available online: https://www.plasticseurope.org/download_file/force/4261/181 (accessed on 16 December 2020).
2. Jambeck, J.R.; Geyer, R.; Wilcox, C.; Siegler, T.R.; Perryman, M.; Andrady, A.; Narayan, R.; Lavender Law, K. Plastic waste inputs from land into the ocean. *Science (80-)* **2015**, *347*, 768–771. [CrossRef]
3. Kosior, E.; Mitchell, J. *Current Industry Position on Plastic Production and Recycling*; Elsevier Inc.: Amsterdam, The Netherlands, 2020; ISBN 9780128178805.
4. Achilias, D.S.; Roupakias, C.; Megalokonomos, P.; Lappas, A.A.; Antonakou, E.V. Chemical recycling of plastic wastes made from polyethylene (LDPE and HDPE) and polypropylene (PP). *J. Hazard. Mater.* **2007**, *149*, 536–542. [CrossRef] [PubMed]

5.  Anuar Sharuddin, S.D.; Abnisa, F.; Wan Daud, W.M.A.; Aroua, M.K. A review on pyrolysis of plastic wastes. *Energy Convers. Manag.* **2016**, *115*, 308–326. [CrossRef]
6.  Chiemchaisri, C.; Charnnok, B.; Visvanathan, C. Recovery of plastic wastes from dumpsite as refuse-derived fuel and its utilization in small gasification system. *Bioresour. Technol.* **2010**, *101*, 1522–1527. [CrossRef]
7.  Tokiwa, Y.; Calabia, B.P. Degradation of microbial polyesters. *Biotechnol. Lett.* **2004**, *26*, 1181–1189. [CrossRef]
8.  Bhardwaj, H.; Gupta, R.; Tiwari, A. Communities of microbial enzymes associated with biodegradation of plastics. *J. Polym. Environ.* **2013**, *21*, 575–579. [CrossRef]
9.  Ahmed, T.; Shahid, M.; Azeem, F.; Rasul, I.; Shah, A.A.; Noman, M.; Hameed, A.; Manzoor, N.; Manzoor, I.; Muhammad, S. Biodegradation of plastics: Current scenario and future prospects for environmental safety. *Environ. Sci. Pollut. Res.* **2018**, *25*, 7287–7298. [CrossRef]
10. Hadad, D.; Geresh, S.; Sivan, A. Biodegradation of polyethylene by the thermophilic bacterium *Brevibacillus borstelensis. J. Appl. Microbiol.* **2005**, *98*, 1093–1100. [CrossRef] [PubMed]
11. Tomita, K.; Kuroki, Y.; Nagai, K. Isolation of thermophiles degrading poly(L-lactic acid). *J. Biosci. Bioeng.* **1999**, *87*, 752–755. [CrossRef]
12. Obradors, N.; Aguilar, J. Efficient biodegradation of high-molecular-weight polyethylene glycols by pure cultures of *Pseudomonas stutzeri. Appl. Environ. Microbiol.* **1991**, *57*, 2383–2388. [CrossRef]
13. Walter, T.; Augusta, J.; Müller, R.J.; Widdecke, H.; Klein, J. Enzymatic degradation of a model polyester by lipase from *Rhizopus delemar. Enzym. Microb. Technol.* **1995**, *17*, 218–224. [CrossRef]
14. Nishide, H.; Toyota, K.; Kimura, M. Effects of soil temperature and anaerobiosis on degradation of biodegradable plastics in soil and their degrading microorganisms. *Soil Sci. Plant Nutr.* **1999**, *45*, 963–972. [CrossRef]
15. Billen, P.; Khalifa, L.; Van Gerven, F.; Tavernier, S.; Spatari, S. Technological application potential of polyethylene and polystyrene biodegradation by macro-organisms such as mealworms and wax moth larvae. *Sci. Total Environ.* **2020**, *735*, 139521. [CrossRef] [PubMed]
16. Bożek, M.; Hanus-Lorenz, B.; Rybak, J. The studies on waste biodegradation by *Tenebrio molitor. E3S Web Conf.* **2017**, *17*, 00011. [CrossRef]
17. Przemieniecki, S.W.; Kosewska, A.; Ciesielski, S.; Kosewska, O. Changes in the gut microbiome and enzymatic profile of *Tenebrio molitor* larvae biodegrading cellulose, polyethylene and polystyrene waste. *Environ. Pollut.* **2020**, *256*, 113265. [CrossRef] [PubMed]
18. Urbanek, A.K.; Rybak, J.; Wróbel, M.; Leluk, K.; Mirończuk, A.M. A comprehensive assessment of microbiome diversity in *Tenebrio molitor* fed with polystyrene waste. *Environ. Pollut.* **2020**, *262*, 114281. [CrossRef] [PubMed]
19. Wu, Q.; Tao, H.; Wong, M.H. Feeding and metabolism effects of three common microplastics on *Tenebrio molitor* L. *Environ. Geochem. Health* **2019**, *41*, 17–26. [CrossRef]
20. Yang, Y.; Yang, J.; Wu, W.M.; Zhao, J.; Song, Y.; Gao, L.; Yang, R.; Jiang, L. Biodegradation and mineralization of polystyrene by plastic-eating mealworms: Chemical and physical characterization and isotopic tests. *Environ. Sci. Technol.* **2015**, *49*, 12080–12086. [CrossRef]
21. Yang, Y.; Wang, J.; Xia, M. Biodegradation and mineralization of polystyrene by plastic-eating superworms *Zophobas atratus. Sci. Total Environ.* **2020**, *708*, 135233. [CrossRef]
22. Aboelkheir, M.G.; Visconte, L.Y.; Oliveira, G.E.; Toledo Filho, R.D.; Souza, F.G. The biodegradative effect of *Tenebrio molitor* Linnaeus larvae on vulcanized SBR and tire crumb. *Sci. Total Environ.* **2019**, *649*, 1075–1082. [CrossRef]
23. Abdulhay, H.S. Biodegradation of plastic wastes by confused flour beetle *Tribolium confusum* Jacquelin du Val larvae. *Asian J. Agric. Biol.* **2020**, *8*, 201–206. [CrossRef]
24. Lou, Y.; Ekaterina, P.; Yang, S.S.; Lu, B.; Liu, B.; Ren, N.; Corvini, P.F.X.; Xing, D. Biodegradation of polyethylene and polystyrene by greater wax moth larvae (*Galleria mellonella* L.) and the effect of co-diet supplementation on the core gut microbiome. *Environ. Sci. Technol.* **2020**, *54*, 2821–2831. [CrossRef]
25. Zhang, J.; Gao, D.; Li, Q.; Zhao, Y.; Li, L.; Lin, H.; Bi, Q.; Zhao, Y. Biodegradation of polyethylene microplastic particles by the fungus *Aspergillus flavus* from the guts of wax moth *Galleria mellonella. Sci. Total Environ.* **2020**, *704*. [CrossRef]
26. Kesti, S.S.; Thimmappa, C.S. First report on biodegradation of low density polyethylene by rice moth larvae, *Corcyra cephalonica* (stainton). *Holist. Approach Environ.* **2019**, *9*, 79–83. [CrossRef]
27. Ewuim, S. Entomoremediation—A novel in-Situ bioremediation approach. *Anim. Res. Int.* **2013**, *10*, 1681–1684. [CrossRef]
28. Bulak, P.; Polakowski, C.; Nowak, K.; Waśko, A.; Wiącek, D.; Bieganowski, A. *Hermetia illucens* as a new and promising species for use in entomoremediation. *Sci. Total Environ.* **2018**, *633*, 912–919. [CrossRef] [PubMed]
29. Song, Y.; Qiu, R.; Hu, J.; Li, X.; Zhang, X.; Chen, Y.; Wu, W.M.; He, D. Biodegradation and disintegration of expanded polystyrene by land snails *Achatina fulica. Sci. Total Environ.* **2020**, *746*, 141289. [CrossRef] [PubMed]
30. Mariod, A.A.; Mirghani, M.E.S.; Hussein, I. *Tenebrio molitor* Mealworm. *Unconv. Oilseeds Oil Sources* **2017**, 331–336. [CrossRef]
31. Finke, M.D. Complete nutrient content of four species of commercially available feeder insects fed enhanced diets during growth. *Zoo Biol.* **2015**, *34*, 554–564. [CrossRef]
32. Jajic, I.; Popovic, A.; Urosevic, M.; Krstovic, S.; Petrovic, M.; Guljas, D.; Samardzic, M. Fatty and amino acid profile of mealworm larvae (*Tenebrio molitor* L.). *Biotechnol. Anim. Husb.* **2020**, *36*, 167–180. [CrossRef]

33. Jansen, Z. The Nutritional Potential of Black Soldier Fly (*Hermetia Illucens*) Larvae for Layer Hens. Ph.D. Thesis, Stellenbosch University, Stellenbosch, South Africa, 2018; pp. 1–103.
34. Rivero Pino, F.; Pérez Gálvez, R.; Espejo Carpio, F.J.; Guadix, E.M. Evaluation of: *Tenebrio molitor* protein as a source of peptides for modulating physiological processes. *Food Funct.* **2020**, *11*, 4376–4386. [CrossRef]
35. Houben, D.; Daoulas, G.; Faucon, M.P.; Dulaurent, A.M. Potential use of mealworm frass as a fertilizer: Impact on crop growth and soil properties. *Sci. Rep.* **2020**, *10*, 4659. [CrossRef] [PubMed]
36. Bulak, P.; Proc, K.; Pawłowska, M.; Kasprzycka, A.; Berus, W.; Bieganowski, A. Biogas generation from insects breeding post production wastes. *J. Clean. Prod.* **2020**, *244*, 531–537. [CrossRef]
37. Kwon, G.T.; Yuk, H.G.; Lee, S.J.; Chung, Y.H.; Jang, H.S.; Yoo, J.S.; Cho, K.H.; Kong, H.; Shin, D. Mealworm larvae (*Tenebrio molitor* L.) exuviae as a novel prebiotic material for BALB/c mouse gut microbiota. *Food Sci. Biotechnol.* **2020**, *29*, 531–537. [CrossRef] [PubMed]
38. Dabbou, S.; Ferrocino, I.; Gasco, L.; Schiavone, A.; Trocino, A.; Xiccato, G.; Barroeta, A.C.; Maione, S.; Soglia, D.; Biasato, I.; et al. Antimicrobial effects of black soldier fly and yellow mealworm fats and their impact on gut microbiota of growing rabbits. *Animals* **2020**, *10*, 1292. [CrossRef] [PubMed]
39. Saenz-Mendoza, A.I.; Zamudio-Flores, P.B.; García-Anaya, M.C.; Velasco, C.R.; Acosta-Muñiz, C.H.; de Jesús Ornelas-Paz, J.; Hernández-González, M.; Vargas-Torres, A.; Aguilar-González, M.Á.; Salgado-Delgado, R. Characterization of insect chitosan films from *Tenebrio molitor* and *Brachystola magna* and its comparison with commercial chitosan of different molecular weights. *Int. J. Biol. Macromol.* **2020**, *160*, 953–963. [CrossRef]
40. Li, X.; Chen, L.; Mei, Q.; Dong, B.; Dai, X.; Ding, G.; Zeng, E.Y. Microplastics in sewage sludge from the wastewater treatment plants in China. *Water Res.* **2018**, *142*, 75–85. [CrossRef]
41. Yang, S.S.; Brandon, A.M.; Andrew Flanagan, J.C.; Yang, J.; Ning, D.; Cai, S.Y.; Fan, H.Q.; Wang, Z.Y.; Ren, J.; Benbow, E.; et al. Biodegradation of polystyrene wastes in yellow mealworms (larvae of *Tenebrio molitor* Linnaeus): Factors affecting biodegradation rates and the ability of polystyrene-fed larvae to complete their life cycle. *Chemosphere* **2018**, *191*, 979–989. [CrossRef]
42. Peng, B.Y.; Su, Y.; Chen, Z.; Chen, J.; Zhou, X.; Benbow, M.E.; Criddle, C.S.; Wu, W.M.; Zhang, Y. Biodegradation of polystyrene by Dark (*Tenebrio obscurus*) and Yellow (*Tenebrio molitor*) Mealworms (Coleoptera: Tenebrionidae). *Environ. Sci. Technol.* **2019**, *53*, 5256–5265. [CrossRef]
43. Brandon, A.M.; Gao, S.H.; Tian, R.; Ning, D.; Yang, S.S.; Zhou, J.; Wu, W.M.; Criddle, C.S. Biodegradation of polyethylene and plastic mixtures in mealworms (larvae of *Tenebrio molitor*) and effects on the gut microbiome. *Environ. Sci. Technol.* **2018**, *52*, 6526–6533. [CrossRef]
44. Guo, B.; Yin, J.; Hao, W.; Jiao, M. Polyurethane foam induces epigenetic modification of mitochondrial DNA during different metamorphic stages of *Tenebrio molitor*. *Ecotoxicol. Environ. Saf.* **2019**, *183*, 109461. [CrossRef]
45. Khan, S.; Nadir, S.; Mortimer, P.E.; Xu, J.; Gui, H.; Khan, A.; Iqbal, S.; Ye, L.; Shi, L. Biodegradation of polyester polyurethane by *Aspergillus tubingensis*. *Environ. Pollut.* **2019**, *229*, 469–480. [CrossRef] [PubMed]
46. Yang, L.; Gao, J.; Liu, Y.; Zhuang, G.; Peng, X.; Wu, W.M.; Zhuang, X. Biodegradation of expanded polystyrene and low-density polyethylene foams in larvae of *Tenebrio molitor* Linnaeus (Coleoptera: Tenebrionidae): Broad versus limited extent depolymerization and microbe-dependence versus independence. *Chemosphere* **2021**, *262*, 127818. [CrossRef] [PubMed]
47. Yang, S.S.; Wu, W.M.; Brandon, A.M.; Fan, H.Q.; Receveur, J.P.; Li, Y.; Wang, Z.Y.; Fan, R.; McClellan, R.L.; Gao, S.H.; et al. Ubiquity of polystyrene digestion and biodegradation within yellow mealworms, larvae of *Tenebrio molitor* Linnaeus (Coleoptera: Tenebrionidae). *Chemosphere* **2018**, *212*, 262–271. [CrossRef] [PubMed]
48. Ranta-Korpi, M.; Vainikka, P.; Konttinen, J.; Saarimaa, A.; Rodriguez, M. *Ash Forming Elements in Plastics and Rubbers*; VTT Technical Research Centre of Finland: Espoo, Finland, 2014; ISBN 978-951-38-8158-0.
49. Turner, A. Heavy metals, metalloids and other hazardous elements in marine plastic litter. *Mar. Pollut. Bull.* **2016**, *111*, 136–142. [CrossRef]
50. Sing, H.; Jain, A. Ignition, combustion, toxicity, and fire retardancy of polyurethane foams: A comprehensive review. *J. Appl. Polym. Sci.* **2008**, *111*, 1115–1143. [CrossRef]
51. Alt, H.G.; Samuel, E. Fluorenyl complexes of zirconium and hafnium as catalysts for olefin polymerization. *Chem. Soc. Rev.* **1998**, *27*, 323–329. [CrossRef]
52. Lee, H.M.; Kim, H.R.; Jeon, E.; Yu, H.C.; Lee, S.; Li, J.; Kim, D.-H. Evaluation of the biodegradation efficiency of four various types of plastics by *Pseudomonas aeruginosa* isolated from the gut extract of superworms. *Microorganisms* **2020**, *8*, 1341. [CrossRef]
53. Masykuri, M.; Widyasari, F. Surface morphology and biodegradation tests of polyurethane/clay rigid foam composites. *J. Chem. Technol. Metall.* **2021**, *56*, 744–753.
54. Ichikawa, T.; Kurauchi, T. Larval cannibalism and pupal defense against cannibalism in two species of tenebrionid beetles. *Zool. Sci.* **2009**, *26*, 525–529. [CrossRef]

Article

# Polyurethane Foam Residue Biodegradation through the *Tenebrio molitor* Digestive Tract: Microbial Communities and Enzymatic Activity

Jose M. Orts [1], Juan Parrado [1,*], Jose A. Pascual [2,*], Angel Orts [1], Jessica Cuartero [2], Manuel Tejada [3] and Margarita Ros [2]

1   Departament of Biochemistry and Molecular Biology, Facultad de Farmacia, Universidad de Sevilla,
    C/Prof. García Gonzalez 2, 41012 Sevilla, Spain
2   Department of Soil and Water Conservation and Organic Waste Management, Centro de Edafologia y Biologia
    Aplicada del Segura (CEBAS-CSIC), University Campus of Espinardo, 30100 Murcia, Spain
3   Grupo de Investigacion Edafologia Ambiental, Departamento de Cristalografia, Mineralogia y Quimica
    Agricola, E.T.S.I.A. Universidad de Sevilla, 41004 Sevilla, Spain
*   Correspondence: parrado@us.es (J.P.); jpascual@cebas.csic.es (J.A.P.)

**Abstract:** Polyurethane (PU) is a widely used polymer with a highly complex recycling process due to its chemical structure. Eliminating polyurethane is limited to incineration or accumulation in landfills. Biodegradation by enzymes and microorganisms has been studied for decades as an effective method of biological decomposition. In this study, *Tenebrio molitor* larvae (*T. molitor*) were fed polyurethane foam. They degraded the polymer by 35% in 17 days, resulting in a 14% weight loss in the mealworms. Changes in the *T. molitor* gut bacterial community and diversity were observed, which may be due to the colonization of the species associated with PU degradation. The physical and structural biodegradation of the PU, as achieved by *T. molitor*, was observed and compared to the characteristics of the original PU (PU-virgin) using Fourier Transform InfraRed spectroscopy (FTIR), Thermal Gravimetric Analysis (TGA), and Scanning Electron Microphotography (SEM).

**Keywords:** plastic; mealworms; insect; bacteria; gut microbiome

**Citation:** Orts, J.M.; Parrado, J.; Pascual, J.A.; Orts, A.; Cuartero, J.; Tejada, M.; Ros, M. Polyurethane Foam Residue Biodegradation through the *Tenebrio molitor* Digestive Tract: Microbial Communities and Enzymatic Activity. *Polymers* **2023**, *15*, 204. https://doi.org/10.3390/polym15010204

Academic Editor: Piotr Bulak

Received: 22 November 2022
Revised: 7 December 2022
Accepted: 27 December 2022
Published: 31 December 2022

## 1. Introduction

Worldwide plastic polymer production and consumption are increasing every year, reaching around 350–400 million tons in 2019. Its reduction and elimination are a huge challenge [1] since an estimated 2.1 to 3.6 million tons are generated per year in Europe [2]. Polyurethanes (PU) are a type of plastic polymer used as foam in automobile seats, coatings, sealants, the textile industry and other areas [3], with around 4 million tons produced worldwide in 2019 [1]. PU foam is synthesized by the reaction of Diisocyanate (R–N=C=O) and Polyol (R'–OH), producing organic units called urethane [4]. Numerous groups of urethane joined together form a polyurethane molecule [5,6]. Polyurethanes are characterized as resistant to physical and biological degradation due to their chemical composition, which is highly resistant to temperatures and hydrophobicity, resulting in their long lifespans [7].

Recycling polyurethane foam (PU foam) is complicated [8], although mechanical transformation processes, such as crushing and compression molding or pulverizing, can be employed [9]. PU foam waste can also be used as a filler load in lower-value products [10]. However, not enough PU foam is used this way, and the large amount of accumulated PU waste is a big environmental problem. This PU waste is usually incinerated, and the gas emitted in this process contributes to the greenhouse effect. Potentially toxic gases are also emitted from polyurethane combustion, and that not burned accumulates in landfills or aquatic systems [8,11].

New approaches to PU biodegradation using enzymes, e.g., cholesterol esterases, proteases, lipases [12,13], or microorganisms, e.g., *Cladosporium pseudocladosporioides* or

*Paracoccus* sp. [7,14] that attack the PU bonds are being investigated, with interesting degradation results. The use of insects to biodegrade plastic is a new natural approach [15,16]. During the last ten years, *Tenebrio molitor* larvae (*T. molitor*; yellow mealworms), commercially used as animal feed and potential alternative protein for human consumption [17,18], have been reported to ingest and degrade different types of plastics, i.e., polyethylene (PE), polystyrene (PS), or the less-studied polyurethane (PU) [15,16,19]. Studies have shown that *T. molitor*'s ability to biodegrade plastics is mainly dependent on its intestinal microorganisms and adaptation to different foods [20,21].

The purpose of this study was to examine: (i) the feasibility of biodegradation, the chemical and physical changes in PU foam using the mealworm larvae of *T. molitor*; and (ii) the gut enzyme activity and microbiota changes associated with feeding polyurethane to *T. molitor*.

## 2. Materials and Methods

### 2.1. Plastics and Mealworms

The polyurethane foam (PU) used in this experiment was obtained from Interplasp S.L. (total Carbon $540.8 \pm 15.6$ g kg$^{-1}$; total Nitrogen $43.6 \pm 1.09$ g kg$^{-1}$; total Phosphorous $0.0037 \pm 0.0016$ g kg$^{-1}$; total Potassium $0.1 \pm 0.02$ g kg$^{-1}$). An Inductively Coupled Plasma (ICP) analysis was carried out to check for the presence of heavy metals in the polyurethane composition that could be toxic to *T. molitor* larvae, such as Cd, Hg, or Pb [22].

The mealworm larvae of *Tenebrio molitor* (*T. molitor*) were purchased from Proteinsecta (Albacete, Murcia). Before starting the experiment, the mealworms were fed wheat bran except for the last 24 h, when all food was withdrawn. They were fed PU or wheat bran (bran) for 17 days, with the food incorporated at the beginning of the experiment and not re-established. Three replicate containers with 100 g of randomly selected mealworms were fed 15 g of PU foam, as the PU diet (PU), and three replicate containers of 100 g of randomly selected mealworms were fed 100 g of wheat bran, as the control diet (bran). The containers were incubated in the dark at $27 \pm 1$ °C and with $80 \pm 3\%$ relative humidity. This made it possible to determine the behavior of the microbiota and metabolism of the mealworms when they did not obtain any additional nutrients.

### 2.2. Analysis of PU Foam Biodegradation

#### 2.2.1. Polyurethane Consumption

To evaluate the PU foam biodegradation, the PU and mealworms were sampled at 3, 10, and 17 days after feeding the mealworms with PU and wheat bran. At each sampling, the mealworms, PU, and frass (feces) were separated and stored. The equivalent of 5 g of mealworms was taken out and frozen for further analysis. The PU foam and the mealworms were weighed to calculate the mealworms' weight loss (%) and PU consumption (%).

#### 2.2.2. Analysis of the PU Foam with Fourier Transform InfraRed Spectroscopy (FTIR)

Fourier Transform InfraRed spectroscopy with attenuated Total Reflectance Analysis (FTIR-ATR) (Bruker Hyperion 1000, Billerica, Massachusetts, USA) in the wave range of 3100–400 cm$^{-1}$ was used to analyze the changes in the bonds of the PU foam (PU) at the end of the experiment (17 days) compared to the original polyurethane foam (PU-virgin). The foam was previously washed with distilled water ($\times 3$) and dried in an oven for 24 h at 80 °C.

#### 2.2.3. Analysis of the PU Foam with Thermogravimetric Analysis (TGA)

This analysis was performed using an SDT Q600 Thermogravimetric analyzer, Waters, TA instruments, Milford Massachusetts (USA) to characterize the changes in the thermal properties of the PU foam (PU) at the end of the experiment (17 days) compared to the original polyurethane foam (PU-virgin). The TGA was performed at 10 °C min$^{-1}$ from 30 °C to 700 °C in a nitrogen atmosphere (flow rate 25 mL min$^{-1}$) in 2–3 mg of polyurethane foam.

2.2.4. Analysis of the PU Foam with Scanning Electron Microphotography (SEM)

The PU foam (PU) and original PU foam (PU-virgin) were analyzed after 10 and 17 days as feed for the mealworms by Scanning Electron Microscopy (SEM) using FEI TENEO New York (USA). The analysis was developed following [14]. Previously, the PU foam was washed by immersion in 0.88% ($w/v$) sodium hypochlorite for 2 h to eliminate any possible remains of microorganisms in the foam. Later, it was washed in triplicate in 100 mL of distilled water, stirred for 2 min at 150 rpm, and dried for 24 h at 80 °C and coated in platinum.

*2.3. Gut Microbiome Analysis*

2.3.1. Enzyme Activities

The enzymatic activity of the esterases, lipases, proteases, and laccases able to break some of the specific bonds that form polyurethane molecules was measured in the gut of the mealworm larvae fed with PU foam and bran after 3, 10, and 17 days. The gut was obtained by dissecting three larvae previously washed with 2 mL of distilled water and air-dried. Once dissected, the gut was immersed in 1 mL of 0.1 M phosphate buffer (pH 7) and shaken. Subsequently, the homogenate was centrifuged (Eppendorf, MiniSpin, Hamburg, Germany) at $14,100 \times g$ for 10 min. The supernatant was diluted 1:10, and an enzyme activity analysis was carried out (sample). All the enzyme activity was analyzed on a GeneQuant 1300 spectrophotometer, VWR, Radnor, PA, USA. *Protease activity* was determined spectrophotometrically by hydrolysis of p-nitroaniline following the modified method of Preiser et al. [23]. A sample of 100 µL was mixed with 700 µL of reaction mixture and incubated for 90 min at 37 °C. The reaction was stopped with 800 µL of 30% ($v/v$) acetic acid, and the color change was measured at $\lambda = 410$ nm. The reaction mixture contained 0.05 M glycine Na–OH buffer (pH 10) and 0.001 M BAPNA dissolved in 1000 µL of DMSO. The control samples were analyzed in the same way, but the sample was replaced with a glycine Na–OH buffer. *Esterase activity* was determined spectrophotometrically by hydrolysis of p-nitrophenyl acetate (p-NPA) following the modified method of Oceguera-Cervantes et al. [24]. In a final volume of 1 mL, 100 µL of sample was mixed with 800 µL of sodium phosphate buffer (0.05 M; pH 6.5) and 100 µL of p-NPA solution in acetonitrile (0.01 M). The samples were incubated at 37 °C for 20 min. The reaction was stopped by placing them in an ice bath for 5 min. Subsequently, the samples were centrifuged at $10,000 \times g$ for 5 min and measured at $\lambda = 410$ nm. The control samples were analyzed in the same way, but the sample was replaced with a sodium phosphate buffer. *Lipase activity* was determined spectrophotometrically with the hydrolysis of p-nitrophenyl laurate (p-NPL) using a modification of the Kilcawley et al. [13] method. In a final volume of 2 mL, 100 µL of sample was mixed with 1.9 mL of sodium phosphate buffer (0.1 M; pH 7). The samples were incubated at 37 °C for 30 min. The reaction was stopped by placing them in an ice bath for 5 min and adding 0.85 mL of NaOH (0.5 M). Subsequently, the samples were centrifuged at $10,000 \times g$ for 5 min and measured at $\lambda = 400$ nm. The control samples were analyzed in the same way, but the sample was replaced with a sodium phosphate buffer. *Laccase activity* was measured using spectrophotometry following the modified method of Dhakar and Pandey [25]. The following were mixed and incubated for 1 min: 1 mL of final volume, 100 µL of sample, 800 µL of sodium acetate buffer (0.2 M/0.1 M; pH 4.5), and 100 µL of ABTS (0.01 M) dissolved in a sodium acetate buffer (0.2 M/0.1 M; pH 4.5). Subsequently, the samples were centrifuged at $10,000 \times g$ for 5 min and measured at $\lambda = 420$ nm. The control samples were analyzed in the same way, but the samples were replaced with a sodium acetate buffer.

$\varepsilon$ The calculations of each activity were carried out using the molar extinction coefficients. Protease activity was calculated using the molar extinction coefficient at $\lambda = 410$ nm by $\varepsilon = 8.8$ mM$^{-1}$ cm$^{-1}$ [26]; esterase activity at $\lambda = 410$ nm by $\varepsilon = 18.5$ mM$^{-1}$ cm$^{-1}$ [27]; lipase activity by $\lambda = 400$ nm $\varepsilon = 14.8$ mM$^{-1}$ cm$^{-1}$ [13], and laccase activity at $\lambda = 420$ nm by $\varepsilon = 36$ mM$^{-1}$ cm$^{-1}$ [25]. All the activities were expressed in µM$^{-1}$ g$^{-1}$ min$^{-1}$.

### 2.3.2. Gut DNA Extraction and Amplicon Sequencing

The DNA from the gut of mealworms fed with PU foam (PU) and wheat bran (bran) for 3, 10, and 17 days was extracted from the gut of four mealworms from the same feed container pooled to eliminate individual variability. The gut was harvested and added to a tube with 100 μL of phosphate buffer (0.1 M). The DNA was extracted using the Dneasy PowerSoil kit (Qiagen, Hilden, Germany). The quantity and quality of the DNA extracts were evaluated using a Qubit 3.0 Fluorometer (Invitrogen, Thermo Fisher Scientific, Waltham, MA, USA). The samples for sequencing were analyzed with the Illumina Nextera barcoded two-step PCR libraries (V4, ITS2) and sequenced on an Illumina MiSeq, v3, $2 \times 300$ bp. Demultiplexing and trimming the Illumina adaptor residuals and trimming the locus specific primer sequences were removed (Microsynth AG, Balgach, Switzerland).

### 2.3.3. Bioinformatic Analysis

The analysis of the sequences was carried out as follows: the samples were received in fastq, entered into QIIME2 [28], and denoised using the dada2 algorithm [29], taking into account the replicates of each treatment to avoid their slight variations. Once the ASV (Amplicon Sequence Variant) was obtained, the taxonomic classification was performed using consensus-vsearch with the Silva 132 database (released in 2020) as a reference. The eukaryotes, archeas, mitochondria, chloroplasts, and non-assignments were eliminated before the statistical analysis. Finally, normalization was conducted through rarefaction, bringing all the samples to the same value using the value of the last sample that conserved a minimum of 3000 readings. The sequences were deposited in the DNA database with the accession code PRJEB54959.

### 2.4. Statistical Analysis

The results were analyzed using Statgraphics 18 and plotted through Sigmaplot v12.0. The correlations were performed using Spearman's correlation test, and the statistical significance was determined using a two-way analysis of variance (ANOVA) with a post-hoc Least Significant Difference (LSD) Fisher Test. To study the microbial community, an NMDS was performed using the Bray–Curtis distance with the vegan package and the graphs were made using the ggplot2 package.

## 3. Results and Discussion

### 3.1. PU Foam Consumption and Its Effect on Mealworms

To the best of our knowledge, there are only a few reports about the efficiency of PU consumption by mealworms. The PU consumption by mealworm larvae was linear, showing a lower slope ($y_1 = 2.995 + 1.893x$) than the bran consumption ($y_2 = 16.424 + 8.049x$) (Figure S1 Supplementary Materials). This reflected a lower PU consumption (35%) than the bran (100%), and a higher mealworm weight loss (86%) than the bran (97%) (Table 1).

**Table 1.** The mealworm larvae's weight loss and feed consumption.

| Days | Mealworm Weight Loss PU Diet (%) | Mealworm Weight Loss Bran Diet (%) | PU Consumption (%) | Bran Consumption (%) |
|---|---|---|---|---|
| 0 | 100 | 100 | 0 | 0 |
| 3 | $97.95 \pm 0.2$ | $102.82 \pm 1.35$ | $8.27 \pm 0.5$ | $41.13 \pm 1.04$ |
| 6 | $95.32 \pm 0.63$ | $106.4 \pm 1.12$ | $14.35 \pm 0.57$ | $63.75 \pm 5.35$ |
| 10 | $93.53 \pm 0.78$ | $105.73 \pm 4.36$ | $22.73 \pm 2.44$ | $97.33 \pm 0.79$ |
| 17 | $85.84 \pm 1.61$ | $97.18 \pm 7.82$ | $34.78 \pm 2.48$ | $99.95 \pm 0.05$ |

This could be because the bran contains enough nutrients for the *T. molitor* mealworms' metabolism [15], while PU, as the only source of carbon, does not provide sufficient nutrients to support growth. This was observed by Lou et al. [30] with polyethylene (PE) and polystyrene (PS). It could also be due to the high energy cost of eliminating the toxic compounds derived from the PU foam degradation [21]. Bulak et al. [22] observed 45% of

PU consumption and 26–28% of mealworm weight loss after 58 days. If our experiment had lasted as long as Bulak et al.'s [22], we would have found similar data although mealworm death or metabolic exhaustion could occur [21]. Yang et al. [31] showed that mealworms fed with polystyrene were still capable of successful pupation, which indirectly proves that digestion and assimilation occurred after 32 days, although they had a lower fat content than those who ingested a conventional diet.

The results showed that PU is not as palatable for mealworms as bran, although they have been shown to decompose it [32]. However, according to Peng et al. [15], a higher degradation of polystyrene was observed with a mix of bran or corn flour and polystyrene.

### 3.2. Evidence of PU Foam Biodegradation by Mealworms

Evidence of changes in the functional groups of the PU used to feed the mealworms compared to the PU-virgin was provided by FTIR analysis at the end of the incubation period (17 days) (Figure 1). The main changes observed were in the peak intensity compared to the spike appearance and disappearance (functional groups). The PU showed less intensity in the spectrum peaks, such as 1090–1099 cm$^{-1}$ (C–O–C bond), 1220–1225 cm$^{-1}$ (C–N bond), 1536 (N–H bond), 1630–1736 (C=O bond), 2867–2916 (CH2), and 3288 cm$^{-1}$ (–OH bond). The only signal where the PU-virgin was higher than the PU was at 2930–2940 cm$^{-1}$, corresponding to the symmetrical and asymmetrical stretching vibrations of the CH bonds in the CH2 groups [33]. This general pattern demonstrates that the PU molecule was occluded, and only the external parts could be degraded by microorganisms and extracellular enzymes [32] without distinguishing the hard segments, such as the free aromatic bond breaking [12] or urethane ester/ether, from the soft ones, such as the plane urethane represented by C–N of N–H [32,34].

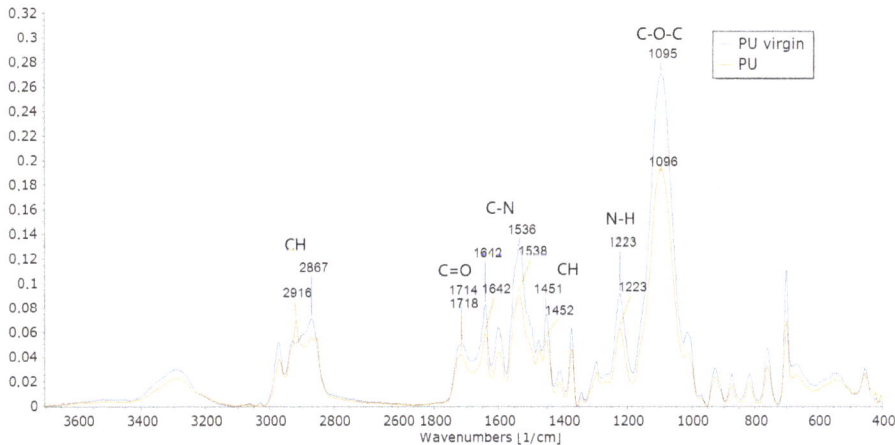

**Figure 1.** The Fourier Transform Infrared spectroscopy (FTIR) analysis of PU foam (PU) and original PU foam (PU-virgin) at the end of the experiment (17 days).

The TGA analysis also provided evidence of modifications in the structure of the PU used to feed the mealworms compared to the PU-virgin after 17 days (Figure 2). The PU weight loss was higher (97.5%) than the PU-virgin (94%). This could be attributed to the lower amount of soft and hard segments found in the PU with higher thermic degradation resistance [7]. The weight loss occurred in three phases: the first stage was from 220 to 280 °C, where about 30% of the weight loss occurred. This could be due to the release of volatile organic compounds [35,36]. The second stage was from 300 to 400 °C, with a weight loss of 60%, corresponding to hard and soft urethane segment dissociation [37]. The third stage was above 400 °C, where a lower fraction of PU (around 10%) was degraded, corresponding to the organic residue decomposition [36] (Figure 2).

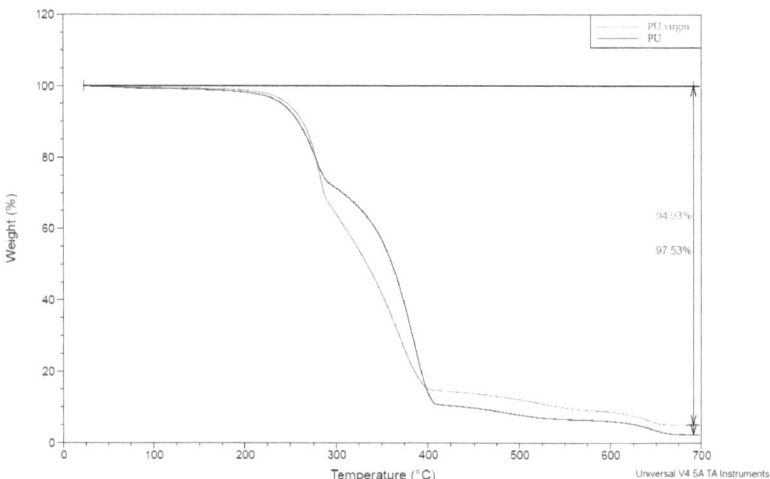

**Figure 2.** The Thermogravimetric analysis (TGA) of the PU foam (PU) and original PU foam (PU-virgin) at the end of the experiment (17 days).

The SEM also demonstrated the physical degradation of the PU after 10 and 17 days compared to the PU-virgin (Figure 3). The surface of the PU-virgin showed smooth edges with no apparent breaks (Figure 3A), while the PU used to feed the mealworms showed wrinkled edges and pits, cracking and erosion that could be attributed to mealworm chewing [32] (Figure 3B–D). Similar results were observed by Bulak et al. [22] and Khan et al. [38] on PU films exposed to *Aspergillus tubingensis*.

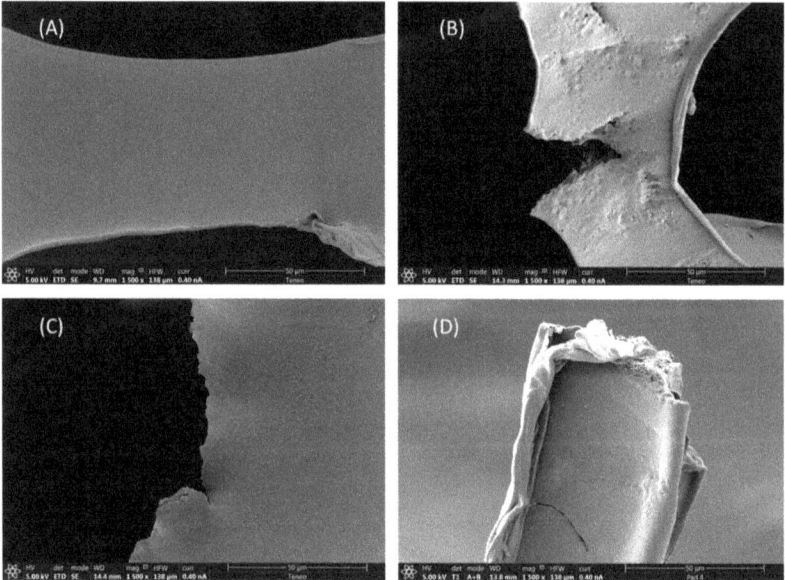

**Figure 3.** Scanning Electron Microphotography (SEM) of different PU foams: PU-virgin (**A**); PU after 10 days (**B**); PU after 17 days (**C,D**).

*3.3. The Mealworm Gut with PU Consumption: The Enzyme Activity and Microbial Community Effect*

3.3.1. Enzyme Activity in the Mealworm Gut

The different enzymatic activity, such as lipases, esterases, proteases, and laccases associated with the polyurethane hydrolysis [39] was measured in the gut of mealworm larvae for both diets, the PU and the bran (Figure 4). These enzymes showed a significant ($p \leq 0.001$) interaction between the type of diet and the sampling time. Enzyme activity in the bran diet was significantly higher than that in the PU diet throughout the experiment, with two apparent phases (Figure 4). Phase one was from 1 to 6 days, with almost constant values, and phase two was from 6 to 17 days, when the enzymatic activity tended to increase. This could be due to the bran depletion from the mealworm consumption. The mealworms thus had fewer available nutrients and a greater need for enzyme synthesis to obtain nutrients from the scarce food available. However, for the PU diet, the behavior was different. The values of the enzyme activity were lower and mostly constant throughout the experiment (Figure 4). This could be explained by the scarce availability of nutrients since the first sampling time, not permitting the synthesis of the required digestive enzymes to degrade the polyurethane [40]. The enzymes esterase and lipase showed higher values (Figure 4A,B) than protease and laccase (Figure 4C,D), probably because the former degraded any PU bond [11]. This low enzyme activity on PU could also be due to the synthesis of the corona protein, which inhibits the absorption of nanoparticles by intestinal cells, such as nanoplastics [41], or the synthesis of enzymes and molecules related to the immune system [42,43].

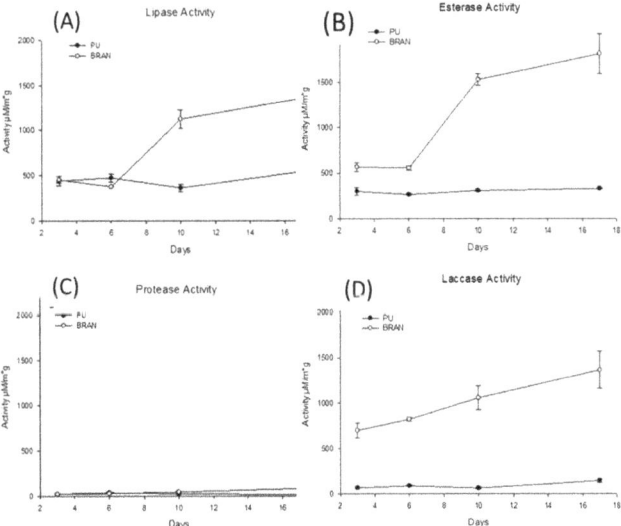

**Figure 4.** Polyurethane activity throughout the experiment for both PU and bran diets (**A**) lipase activity; (**B**) esterase activity; (**C**) protease activity; (**D**) laccase activity. For all the enzyme activity, a Two-Way ANOVA (Pdiet < 0.001); (Ptime < 0.001) was performed.

It has previously been demonstrated than microbiota and the secreted enzymes in the mealworm gut could be adapted to new diets, even poor ones with low nutrient availability, like the PU diet [44]. It has also been shown that one of the survival mechanisms of some insects in situations of stress or nutritional deficit is to consume their lipid reserves [45]. In addition, an increase in proteases inside insects only occurs as a last resort in extremely stressful situations since this would lead to protein biodegradation, which is the last element to degrade [46].

### 3.3.2. The Mealworm Gut Microbial Community

The Illumina MiSeq analysis of PCR-amplified 16S rRNA fragments was used to assess the changes in the gut microbiome community of the mealworms fed PU throughout the experiment since the gut microbiome of insects has an important role to play in their digestion process [47]. The mealworm gut microbiome diversity for the PU diet was higher (average 3.23) than for the bran diet (average 2.90) (Figure 5). Similar results were observed by Wang et al. [32] and Peng et al. [15]. The Shannon diversity for the PU diet slightly increased throughout the experiment, probably due to changes in the proportion of microorganisms capable of degrading PU, while for the bran diet, it slightly decreased.

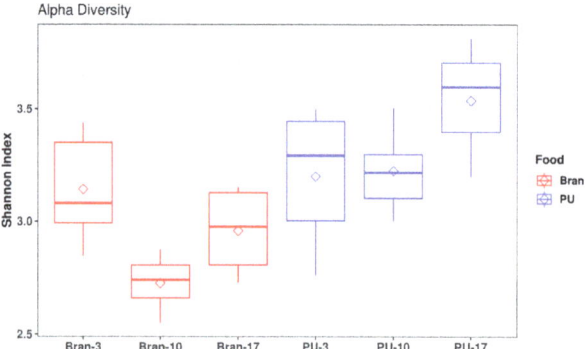

**Figure 5.** The Shannon diversity index of the gut microbiome of the PU and bran diets throughout the sampling (3, 10, and 17 days).

A non-metrical multidimensional scaling (NMDs) exhibited differences in the gut microbiome for the PU diet and the bran diet mainly after 10 and 17 days (Figure 6). The principal phylum observed for both diets (PU and bran) were Tenericutes, Protobacteria, and Firmicutes (Figure 7). Similar results have been found by other authors with different types of plastics [15–17]. Tenericutes was the dominant phylum, reaching the highest proportion for the bran diet (63%), while for the PU diet, it reached 51%. These proportions decreased throughout the experiment. At the end of the experiment (17 days), the opposite occurred, with Tenericutes higher for the PU diet (around 30%) than for the bran diet (17%). Nevertheless, the relative abundance of Firmicutes increased by 18% for the PU diet compared to 10% for the bran diet at the end of the experiment (17 days). Bacteroidetes were lower for the PU diet [48].

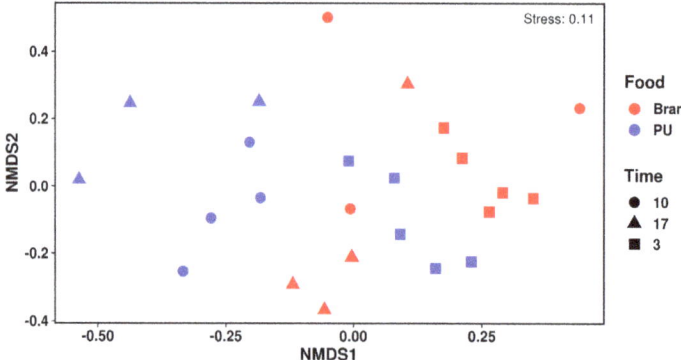

**Figure 6.** The NMDs of the gut microbial community for both diets (PU and bran) at different sampling times (3, 10, and 17 days).

An analysis of the relative abundance at the genus level (Figure 7) indicated four dominant genera for both diets: *Spiroplasma* (average 66.5%), *Enterococcus* (11.08%), *Lactococcus* (9.75%), and *Pediococcus* (5.32%). *Spiroplasma* (Tenericutes) decreased throughout the experiment for the bran diet (7%) and for the PU diet (44%). At the end of the experiment, it was more abundant for the bran diet (69.49%) than for the PU diet (38.62%). *Spiroplasma* was also observed in many studies with different types of plastics [16,20] and it is considered a pathogen or a male-killing bacterium, but in the gut of mealworms, it is not harmful [49].

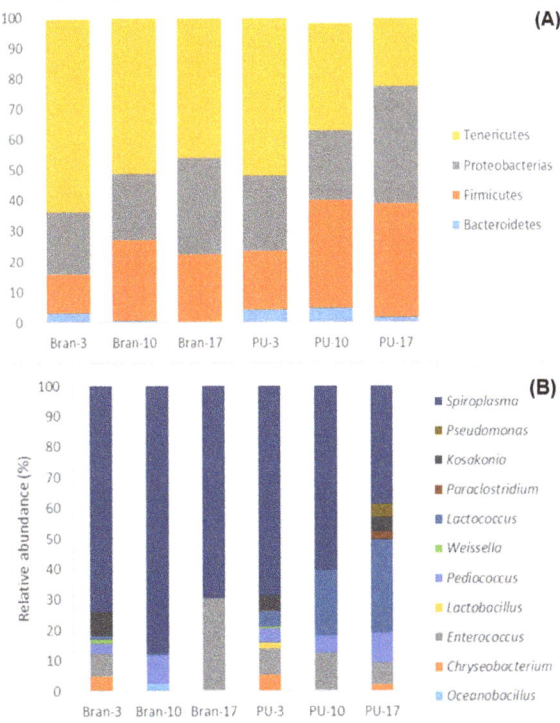

**Figure 7.** A bar plot of the community analysis of the mealworm gut for the PU diet (PU) and the bran diet (bran) with phylum (**A**) and genus (>1%) (**B**).

*Lactococcus* was the only bacteria that could be associated with the PU diet throughout the whole experiment (Figure 8). *Lactococcus* and *Pediococcus* (Firmicutes) (associated with the PU diet at 3–17 days) are lactic acid bacteria, which may have contributed to adjusting and maintaining the health of the gut microbiome [50]. *Lactococcus* and *Enterococcus* (Firmicutes) (associated with the PU diet at 3–17 days) (Figure 8) are also common insect gut bacteria and are known members of the *T. molitor* gut microbiome [20,51]. According to Lou et al. [30], understanding the approximate locations of different bacteria (*Lactococcus* was present in every part of the gut, while *Enterococcus* was absent in the foregut and anterior midgut [51]) is a good way to infer possible degradation pathways in the mealworm gut since *Enterococcus* is related to PU degradation [32].

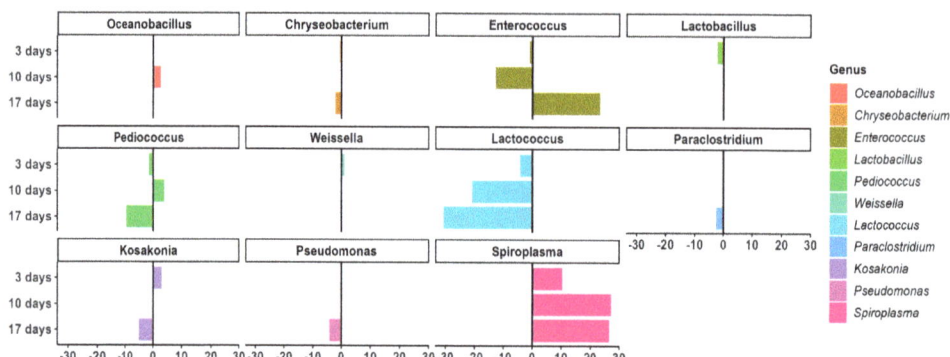

**Figure 8.** The levels and the differential abundance analysis of gut microorganisms for both diets.

At the end of the experiment (17 days), in addition to *Lactococcus, Pediococcus,* and *Enterococcus,* different bacteria could also be associated with the PU diet (Figure 8) such as *Paraclostridium* (Firmicutes), *Chryseobacterium* (Bacteroidetes), *Kosakonia,* and *Pseudomonas* (Proteobacteria). *Chryseobacterium* was observed in the mealworm gut with the different PS diets [52]. *Kosakonia,* a member of the Enterobacteriaceae family has been observed in PE and PS degradation [53–55]. *Pseudomonas* has also been associated with PS biodegradation [16,56]. The digestion process in the intestine of mealworms is more complex than it seems, and the role of whole microbiota and synergic interactions are important in PU degradation [16].

## 4. Conclusions

From our study, we can conclude that the larvae of *T. molitor* were able to consume around 35% of the PU. Larvae chewing increased the surface PU facilitating the mealworm gut microbiome and extracellular enzymes on the PU bond. *Lactococcus* was the only bacteria that could be associated with the PU diet throughout the whole experiment, although other microorganisms, such as *Paraclostridium* (Firmicutes), *Chryseobacterium* (Bacteroidetes), *Kosakonia,* and *Pseudomonas* (Proteobacteria) could also be associated with PU degradation. The above-mentioned PU biodegradation was demonstrated through structural and physical approaches such as FTIR, TGA, and SEM analysis.

**Supplementary Materials:** The following supporting information can be downloaded at: https://www.mdpi.com/article/10.3390/polym15010204/s1, Figure S1: Spearman Correlation between PU and Bran consumption and days of the experiment.

**Author Contributions:** J.A.P. and M.R. carried out the experiments for this study. J.M.O.; M.T. and J.P. performed the FTIR, TAG, and SEM analyses to demonstrate the PU biodegradation evidence. J.M.O. and A.O. also carried out the enzymatic assay and M.R. and J.C. performed the microbial community assay. J.M.O. and M.R. wrote the manuscript with contributions from J.A.P. and J.P. All authors have read and agreed to the published version of the manuscript.

**Funding:** This work was funded by Proyectos Estratégicos Ris3mur (Biomerger Project 2i18sae00058) and Proyectos Estratégicos Orientados a la Transición Ecológica y a la Transición Digital (BIO4Purt project TED2021-131894B-C21).

**Institutional Review Board Statement:** Not applicable.

**Informed Consent Statement:** Not applicable.

**Data Availability Statement:** The data presented in this study are available on request from the corresponding author.

**Acknowledgments:** We are enormously grateful to Emilia Sanz-Rios for her collaboration in the development of the article, as well as for the language corrections.

**Conflicts of Interest:** The authors declare no conflict of interest.

## References

1. Danso, D.; Chow, J.; Streita, W.R. Plastics: Environmental and Biotechnological Perspectives on Microbial Degradation. *Appl. Environ. Microbiol.* **2019**, *85*, 1–14. [CrossRef] [PubMed]
2. Cregut, M.; Bedas, M.; Durand, M.J.; Thouand, G. New Insights into Polyurethane Biodegradation and Realistic Prospects for the Development of a Sustainable Waste Recycling Process. *Biotechnol. Adv.* **2013**, *31*, 1634–1647. [CrossRef] [PubMed]
3. Rowe, L.; Howard, G.T. Growth of Bacillus Subtilis on Polyurethane and the Purification and Characterization of a Polyurethanase-Lipase Enzyme. *Int. Biodeterior. Biodegrad.* **2002**, *50*, 33–40. [CrossRef]
4. Eriksen, M.K.; Pivnenko, K.; Faraca, G.; Boldrin, A.; Astrup, T.F. Dynamic material flow analysis of PET, PE, and PP flows in Europe: Evaluation of the potential for circular economy. *Environ. Sci. Technol.* **2020**, *54*, 16166–16175. [CrossRef] [PubMed]
5. Xie, F.; Zhang, T.; Bryant, P.; Kurusingal, V.; Colwell, J.M.; Laycock, B. Degradation and Stabilization of Polyurethane Elastomers. *Prog. Polym. Sci.* **2019**, *90*, 211–268. [CrossRef]
6. Mahajan, N.; Gupta, P. New Insights into the Microbial Degradation of Polyurethanes. *RSC Adv.* **2015**, *5*, 41839–41854. [CrossRef]
7. Gaytán, I.; Sánchez-Reyes, A.; Burelo, M.; Vargas-Suárez, M.; Liachko, I.; Press, M.; Sullivan, S.; Cruz-Gómez, M.J.; Loza-Tavera, H. Degradation of Recalcitrant Polyurethane and Xenobiotic Additives by a Selected Landfill Microbial Community and Its Biodegradative Potential Revealed by Proximity Ligation-Based Metagenomic Analysis. *Front. Microbiol.* **2020**, *10*, 2986. [CrossRef]
8. Matsumiya, Y.; Murata, N.; Tanabe, E.; Kubota, K.; Kubo, M. Isolation and Characterization of an Ether-Type Polyurethane-Degrading Micro-Organism and Analysis of Degradation Mechanism by *Alternaria* Sp. *J. Appl. Microbiol.* **2010**, *108*, 1946–1953. [CrossRef]
9. Skleničková, K.; Abbrent, S.; Halecký, M.; Kočí, V.; Beneš, H. Biodegradability and Ecotoxicity of Polyurethane Foams: A Review. *Crit. Rev. Environ. Sci. Technol.* **2022**, *52*, 157–202. [CrossRef]
10. Gama, N.V.; Ferreira, A.; Barros-Timmons, A. Polyurethane Foams: Past, Present, and Future. *Materials* **2018**, *11*, 1841. [CrossRef]
11. Liu, J.; He, J.; Xue, R.; Xu, B.; Qian, X.; Xin, F.; Blank, L.M.; Zhou, J.; Wei, R.; Dong, W.; et al. Biodegradation and Up-Cycling of Polyurethanes: Progress, Challenges, and Prospects. *Biotechnol. Adv.* **2021**, *48*, 107730. [CrossRef]
12. Christenson, E.M.; Patel, S.; Anderson, J.M.; Hiltner, A. Enzymatic Degradation of Poly(Ether Urethane) and Poly(Carbonate Urethane) by Cholesterol Esterase. *Biomaterials* **2006**, *27*, 3920–3926. [CrossRef] [PubMed]
13. Kilcawley, K.N.; Wilkinson, M.G.; Fox, P.F. Determination of Key Enzyme Activities in Commercial Peptidase and Lipase Preparations from Microbial or Animal Sources. *Enzyme Microb. Technol.* **2002**, *31*, 310–320. [CrossRef]
14. Álvarez-Barragán, J.; Domínguez-Malfavón, L.; Vargas-Suárez, M.; González-Hernández, R.; Aguilar-Osorio, G.; Loza-Tavera, H. Biodegradative Activities of Selected Environmental Fungi on a Polyester Polyurethane Varnish and Polyether Polyurethane Foams. *Appl. Environ. Microbiol.* **2016**, *82*, 5225–5235. [CrossRef] [PubMed]
15. Peng, B.Y.; Su, Y.; Chen, Z.; Chen, J.; Zhou, X.; Benbow, M.E.; Criddle, C.S.; Wu, W.M.; Zhang, Y. Biodegradation of Polystyrene by Dark (*Tenebrio obscurus*) and Yellow (*Tenebrio molitor*) Mealworms (Coleoptera: Tenebrionidae). *Environ. Sci. Technol.* **2019**, *53*, 5256–5265. [CrossRef] [PubMed]
16. Urbanek, A.K.; Rybak, J.; Wróbel, M.; Leluk, K.; Mirończuk, A.M. A Comprehensive Assessment of Microbiome Diversity in Tenebrio Molitor Fed with Polystyrene Waste. *Environ. Pollut.* **2020**, *262*, 114281. [CrossRef] [PubMed]
17. Przemieniecki, S.W.; Kosewska, A.; Ciesielski, S.; Kosewska, O. Changes in the Gut Microbiome and Enzymatic Profile of Tenebrio Molitor Larvae Biodegrading Cellulose, Polyethylene and Polystyrene Waste. *Environ. Pollut.* **2020**, *256*, 113265. [CrossRef]
18. Borremans, A.; Smets, R.; Van Campenhout, L. Fermentation Versus Meat Preservatives to Extend the Shelf Life of Mealworm (*Tenebrio molitor*) Paste for Feed and Food Applications. *Front. Microbiol.* **2020**, *11*, 1–8. [CrossRef]
19. Bombelli, P.; Howe, C.J.; Bertocchini, F. Polyethylene Bio-Degradation by Caterpillars of the Wax Moth *Galleria mellonella*. *Curr. Biol.* **2017**, *27*, R292–R293. [CrossRef]
20. Brandon, A.M.; Gao, S.H.; Tian, R.; Ning, D.; Yang, S.S.; Zhou, J.; Wu, W.M.; Criddle, C.S. Biodegradation of Polyethylene and Plastic Mixtures in Mealworms (Larvae of *Tenebrio molitor*) and Effects on the Gut Microbiome. *Environ. Sci. Technol.* **2018**, *52*, 6526–6533. [CrossRef]
21. Sanchez-Hernandez, J.C. A Toxicological Perspective of Plastic Biodegradation by Insect Larvae. *Comp. Biochem. Physiol. Part-C Toxicol. Pharmacol.* **2021**, *248*, 109117. [CrossRef] [PubMed]
22. Bulak, P.; Proc, K.; Pytlak, A.; Puszka, A.; Gawdzik, B.; Bieganowski, A. Biodegradation of Different Types of Plastics by *Tenebrio molitor* Insect. *Polymers* **2021**, *13*, 3508. [CrossRef]
23. Preiser, H.; Schmitz, J.; Maestracci, D.; Crane, R.K. Modification of an Assay for Trypsin and Its Application for the Estimation of Enteropeptidase. *Clin. Chim. Acta* **1975**, *59*, 169–175. [CrossRef] [PubMed]
24. Oceguera-Cervantes, A.; Carrillo-García, A.; López, N.; Bolaños-Nuñez, S.; Cruz-Gómez, M.J.; Wacher, C.; Loza-Tavera, H. Characterization of the Polyurethanolytic Activity of Two *Alicycliphilus* Sp. Strains Able to Degrade Polyurethane and N-Methylpyrrolidone. *Appl. Environ. Microbiol.* **2007**, *73*, 6214–6223. [CrossRef]
25. Dhakar, K.; Pandey, A. Laccase Production from a Temperature and PH Tolerant Fungal Strain of Trametes Hirsuta (MTCC 11397). *Enzyme Res.* **2013**, *2013*, 869062. [CrossRef] [PubMed]

26. dos Santos, C.W.V.; da Costa Marques, M.E.; de Araújo Tenório, H.; de Miranda, E.C.; Vieira Pereira, H.J. Purification and Characterization of Trypsin from Luphiosilurus Alexandri Pyloric Cecum. *Biochem. Biophys. Rep.* **2016**, *8*, 29–33. [CrossRef]
27. Deshpande, M.V.; Eriksson, K.E.; Göran Pettersson, L. An Assay for Selective Determination of Exo-1,4,-β-Glucanases in a Mixture of Cellulolytic Enzymes. *Anal. Biochem.* **1984**, *138*, 481–487. [CrossRef]
28. Bolyen, E.; Rideout, J.R.; Dillon, M.R.; Bokulich, N.A.; Abnet, C.C.; Al-Ghalith, G.A.; Alexander, H.; Alm, E.J.; Arumugam, M.; Asnicar, F.; et al. Reproducible, Interactive, Scalable and Extensible Microbiome Data Science Using QIIME 2. *Nat. Biotechnol.* **2019**, *37*, 852–857. [CrossRef]
29. Callahan, B.J.; McMurdie, P.J.; Rosen, M.J.; Han, A.W.; Johnson, A.J.A.; Holmes, S.P. DADA2: High-Resolution Sample Inference from Illumina Amplicon Data. *Nat. Methods* **2016**, *13*, 581–583. [CrossRef]
30. Lou, Y.; Li, Y.; Lu, B.; Liu, Q.; Yang, S.S.; Liu, B.; Ren, N.; Wu, W.M.; Xing, D. Response of the Yellow Mealworm (*Tenebrio molitor*) Gut Microbiome to Diet Shifts during Polystyrene and Polyethylene Biodegradation. *J. Hazard. Mater.* **2021**, *416*. [CrossRef]
31. Yang, S.S.; Wu, W.M.; Brandon, A.M.; Fan, H.Q.; Receveur, J.P.; Li, Y.; Wang, Z.Y.; Fan, R.; McClellan, R.L.; Gao, S.H.; et al. Ubiquity of Polystyrene Digestion and Biodegradation within Yellow Mealworms, Larvae of *Tenebrio molitor* Linnaeus (Coleoptera: Tenebrionidae). *Chemosphere* **2018**, *212*, 262–271. [CrossRef]
32. Wang, Y.; Luo, L.; Li, X.; Wang, J.; Wang, H.; Chen, C.; Guo, H.; Han, T.; Zhou, A.; Zhao, X. Different Plastics Ingestion Preferences and Efficiencies of Superworm (*Zophobas atratus* Fab.) and Yellow Mealworm (*Tenebrio molitor* Linn.) Associated with Distinct Gut Microbiome Changes. *Sci. Total Environ.* **2022**, *837*, 155719. [CrossRef]
33. Marchant, R.E.; Zhao, Q.; Anderson, J.M.; Hiltner, A. Degradation of a Poly(Ether Urethane Urea) Elastomer: Infra-Red and XPS Studies. *Polymer* **1987**, *28*, 2032–2039. [CrossRef]
34. Spontón, M.; Casis, N.; Mazo, P.; Raud, B.; Simonetta, A.; Ríos, L.; Estenoz, D. Biodegradation Study by Pseudomonas Sp. of Flexible Polyurethane Foams Derived from Castor Oil. *Int. Biodeterior. Biodegrad.* **2013**, *85*, 85–94. [CrossRef]
35. Zieleniewska, M.; Leszczyński, M.K.; Kurańska, M.; Prociak, A.; Szczepkowski, L.; Krzyzowska, M.; Ryszkowska, J. Preparation and Characterisation of Rigid Polyurethane Foams Using a Rapeseed Oil-Based Polyol. *Ind. Crops Prod.* **2015**, *74*, 887–897. [CrossRef]
36. Ng, W.S.; Lee, C.S.; Chuah, C.H.; Cheng, S.F. Preparation and Modification of Water-Blown Porous Biodegradable Polyurethane Foams with Palm Oil-Based Polyester Polyol. *Ind. Crops Prod.* **2017**, *97*, 65–78. [CrossRef]
37. Pillai, P.K.S.; Li, S.; Bouzidi, L.; Narine, S.S. Metathesized Palm Oil: Fractionation Strategies for Improving Functional Properties of Lipid-Based Polyols and Derived Polyurethane Foams. *Ind. Crops Prod.* **2016**, *84*, 273–283. [CrossRef]
38. Khan, S.; Nadir, S.; Shah, Z.U.; Shah, A.A.; Karunarathna, S.C.; Xu, J.; Khan, A.; Munir, S.; Hasan, F. Biodegradation of Polyester Polyurethane by Aspergillus Tubingensis. *Environ. Pollut.* **2017**, *225*, 469–480. [CrossRef]
39. Barratt, S.R.; Ennos, A.R.; Greenhalgh, M.; Robson, G.D.; Handley, P.S. Fungi Are the Predominant Micro-Organisms Responsible for Degradation of Soil-Buried Polyester Polyurethane over a Range of Soil Water Holding Capacities. *J. Appl. Microbiol.* **2003**, *95*, 78–85. [CrossRef]
40. Terra, W.R.; Barroso, I.G.; Dias, R.O.; Ferreira, C. *Molecular Physiology of Insect Midgut*, 1st ed.; Elsevier Ltd.: Amsterdam, The Netherlands, 2019; Volume 56, ISBN 9780081028421.
41. Peng, Q.; Liu, J.; Zhang, T.; Zhang, T.X.; Zhang, C.L.; Mu, H. Digestive Enzyme Corona Formed in the Gastrointestinal Tract and Its Impact on Epithelial Cell Uptake of Nanoparticles. *Biomacromolecules* **2019**, *20*, 1789–1797. [CrossRef]
42. Hirt, N.; Body-Malapel, M. Immunotoxicity and Intestinal Effects of Nano- and Microplastics: A Review of the Literature. *Part. Fibre Toxicol.* **2020**, *17*, 1–22. [CrossRef] [PubMed]
43. Sharifinia, M.; Bahmanbeigloo, Z.A.; Keshavarzifard, M.; Khanjani, M.H.; Lyons, B.P. Microplastic Pollution as a Grand Challenge in Marine Research: A Closer Look at Their Adverse Impacts on the Immune and Reproductive Systems. *Ecotoxicol. Environ. Saf.* **2020**, *204*, 111109. [CrossRef] [PubMed]
44. Genta, F.A.; Dillon, R.J.; Terra, W.R.; Ferreira, C. Potential Role for Gut Microbiota in Cell Wall Digestion and Glucoside Detoxification in Tenebrio Molitor Larvae. *J. Insect Physiol.* **2006**, *52*, 593–601. [CrossRef] [PubMed]
45. Matyja, K.; Rybak, J.; Hanus-Lorenz, B.; Wróbel, M.; Rutkowski, R. Effects of Polystyrene Diet on Tenebrio Molitor Larval Growth, Development and Survival: Dynamic Energy Budget (DEB) Model Analysis. *Environ. Pollut.* **2020**, *264*. [CrossRef] [PubMed]
46. Wilkins, R.M. Insecticide Resistance and Intracellular Proteases. *Pest Manag. Sci.* **2017**, *73*, 2403–2412. [CrossRef]
47. Wang, Y.; Zhang, Y. Investigation of Gut-Associated Bacteria in *Tenebrio molitor* (Coleoptera: Tenebrionidae) Larvae Using Culture-Dependent and DGGE Methods. *Ann. Entomol. Soc. Am.* **2015**, *108*, 941–949. [CrossRef]
48. Garofalo, C.; Osimani, A.; Milanović, V.; Taccari, M.; Cardinali, F.; Aquilanti, L.; Riolo, P.; Ruschioni, S.; Isidoro, N.; Clementi, F. The Microbiota of Marketed Processed Edible Insects as Revealed by High-Throughput Sequencing. *Food Microbiol.* **2017**, *62*, 15–22. [CrossRef]
49. Jung, J.; Heo, A.; Woo Park, Y.; Ji Kim, Y.; Koh, H.; Park, W. Gut Microbiota of *Tenebrio molitor* and Their Response to Environmental Change. *J. Microbiol. Biotechnol.* **2014**, *24*, 888–897. [CrossRef]
50. Yan, F.; Polk, D.B. Probiotics and Immune Health. *Curr. Opin. Gastroenterol.* **2011**, *27*, 496–501. [CrossRef]
51. Engel, P.; Moran, N.A. The Gut Microbiota of Insects—Diversity in Structure and Function. *FEMS Microbiol. Rev.* **2013**, *37*, 699–735. [CrossRef]

52. Tsochatzis, E.; Berggreen, I.E.; Tedeschi, F.; Ntrallou, K.; Gika, H.; Corredig, M. Gut Microbiome and Degradation Product Formation during Biodegradation of Expanded Polystyrene by Mealworm Larvae under Different Feeding Strategies. *Molecules* **2021**, *26*, 7568. [CrossRef] [PubMed]
53. Gu, C.T.; Li, C.Y.; Yang, L.J.; Huo, G.C. *Enterobacter xiangfangensis* Sp. Nov., Isolated from Chinese Traditional Sourdough, and Reclassification of *Enterobacter sacchari* Zhu et Al. 2013 as *Kosakonia sacchari* Comb. Nov. *Int. J. Syst. Evol. Microbiol.* **2014**, *64*, 2650–2656. [CrossRef]
54. Gautam, R.; Bassi, A.S.; Yanful, E.K. Candida Rugosa Lipase-Catalyzed Polyurethane Degradation in Aqueous Medium. *Biotechnol. Lett.* **2007**, *29*, 1081–1086. [CrossRef] [PubMed]
55. Shah, A.A.; Hasan, F.; Akhter, J.I.; Hameed, A.; Ahmed, S. Degradation of Polyurethane by Novel Bacterial Consortium Isolated from Soil. *Ann. Microbiol.* **2008**, *58*, 381–386. [CrossRef]
56. Stern, R.V.; Howard, G.T. The Polyester Polyurethanase Gene (*PueA*) from *Pseudomonas chlororaphis* Encodes a Lipase. *FEMS Microbiol. Lett.* **2000**, *185*, 163–168. [CrossRef] [PubMed]

MDPI

Article

# Optimization of Polystyrene Biodegradation by *Bacillus cereus* and *Pseudomonas alcaligenes* Using Full Factorial Design

Martina Miloloža, Šime Ukić ⬤, Matija Cvetnić, Tomislav Bolanča and Dajana Kučić Grgić *⬤

Faculty of Chemical Engineering and Technology, University of Zagreb, Trg Marka Marulića 19, 10 000 Zagreb, Croatia
* Correspondence: dkucic@fkit.hr; Tel.: +385-1-4597-238

**Abstract:** Microplastics (MP) are a global environmental problem because they persist in the environment for long periods of time and negatively impact aquatic organisms. Possible solutions for removing MP from the environment include biological processes such as bioremediation, which uses microorganisms to remove contaminants. This study investigated the biodegradation of polystyrene (PS) by two bacteria, *Bacillus cereus* and *Pseudomonas alcaligenes*, isolated from environmental samples in which MPs particles were present. First, determining significant factors affecting the biodegradation of MP-PS was conducted using the Taguchi design. Then, according to preliminary experiments, the optimal conditions for biodegradation were determined by a full factorial design (main experiments). The RSM methodology was applied, and statistical analysis of the obtained models was performed to analyze the influence of the studied factors. The most important factors for MP-PS biodegradation by *Bacillus cereus* were agitation speed, concentration, and size of PS, while agitation speed, size of PS, and optical density influenced the process by *Pseudomonas alcaligenes*. However, the optimal conditions for biodegradation of MP-PS by *Bacillus cereus* were achieved at $\gamma_{MP} = 66.20$, MP size = 413.29, and agitation speed = 100.45. The best conditions for MP-PS biodegradation by *Pseudomonas alcaligenes* were 161.08, 334.73, and 0.35, as agitation speed, MP size, and OD, respectively. In order to get a better insight into the process, the following analyzes were carried out. Changes in CFU, TOC, and TIC concentrations were observed during the biodegradation process. The increase in TOC values was explained by the detection of released additives from PS particles by LC-MS analysis. At the end of the process, the toxicity of the filtrate was determined, and the surface area of the particles was characterized by FTIR-ATR spectroscopy. Ecotoxicity results showed that the filtrate was toxic, indicating the presence of decomposition by-products. In both FTIR spectra, a characteristic weak peak at 1715 cm$^{-1}$ was detected, indicating the formation of carbonyl groups (−C=O), confirming that a biodegradation process had taken place.

**Keywords:** microplastics; polystyrene; *Bacillus cereus*; *Pseudomonas alcaligenes*; Taguchi design; biodegradation; full factorial design

**Citation:** Miloloža, M.; Ukić, Š.; Cvetnić, M.; Bolanča, T.; Kučić Grgić, D. Optimization of Polystyrene Biodegradation by *Bacillus cereus* and *Pseudomonas alcaligenes* Using Full Factorial Design. *Polymers* **2022**, *14*, 4299. https://doi.org/10.3390/polym14204299

Academic Editor: Piotr Bulak

Received: 1 September 2022
Accepted: 10 October 2022
Published: 13 October 2022

**Publisher's Note:** MDPI stays neutral with regard to jurisdictional claims in published maps and institutional affiliations.

## 1. Introduction

It can be said that we live in the plastic age due to the extensive consumption of plastic products in everyday life. Global plastic production has been increasing since the 1950s, reaching 367 million tons in 2020 [1]. The most frequently produced are polyethylene (PE), polyvinyl chloride (PVC), polypropylene (PP), polystyrene (PS), and polyethylene terephthalate (PET) [2,3]. PS is widely used due to excellent properties such as good mechanical properties, lightweight, versatility, durability, stability, and low cost [4]. It is applicable for food and laboratory containers, disposable plastic accessories, CD/cassette boxes, toys, the automobile industry, and construction materials [5,6]. PS production reached 10.4 million tons in 2018, and it is estimated that global PS production will reach 10.9 million tons in 2024 [7]. During the manufacturing process, various additives such as UV stabilizers, antioxidants, flame retardants, and/or lubricants [4,8] are usually added to PS to improve the properties and applicability of the polymer [9]. Generally, stabilizers and antioxidants are added in an amount range between 0.05 and 3% ($w/w$), depending on the structure of the additive and the polymer. For example, phenolic antioxidants are added in small amounts in high-density polyethylene (HDPE), while phosphites are used in large amounts in high-impact PS (HIPS). In addition, dyes (e.g., azo dyes) are commonly used in PS to provide a bright, transparent color [9], but these compounds have a high migration potential. In addition, carcinogenic polybrominated diphenyl ethers [10,11], styrene oligomers as unintentional additives in expanded polystyrene (EPS) [12], and styrene monomers [13] can also migrate from PS products. This makes the leaching of additives from polymers a potential environmental risk to aquatic organisms. Accordingly, it is not surprising that accumulated MP-PS in the environment lead to changes in the ecosystem. Natural degradation of synthetic thermoplastic PS is a very slow process, but microorganisms can use it as a carbon source [4]. According to Auta et al. [14], PS pollutants persist under technical conditions (biodegradation by bacterial genera of *Bacillus*) for 363.16 (*Bacillus cereus*) and 460.00 days (*Bacillus gottheilii*). The environmental lifespan of PS ranges from 50 to 80 years [15]. Kim et al. [16] and Zhang et al. [17] reported that biodegradation of PS in natural ecosystems is slow and requires several hundred years for complete degradation. However, insect-based systems suggest that a much shorter period of biodegradation of PS can occur within a few weeks.

Bioremediation, as an economical, efficient, and environmentally friendly biological process [18], involves the use of microorganisms, bacteria, mold, and yeast [19] for the purpose of removing pollutants from the environment. Biodegradation processes influence many factors such as abiotic (pH-value, concentration of dissolved oxygen, moisture content, temperature, salinity, and presence or absence of UV radiation), biotic factors (type and number of microorganisms, extracellular enzymes, and biosurfactants), and properties of MP particles (structure, molecular weight, hydrophobicity, functional groups, etc.) [20]. Biodegradation of PS is complex, and it is considered that biodegradation begins after microorganisms colonize PS particles. Thereafter, PS may be degraded into smaller fragments such as oligomers, dimers, and/or monomers. Assimilation can occur due to the ability of microorganisms to use styrene as a carbon source [21]. Pathways for the biodegradation of PS by various enzymes (hydroquinone [22], oxidoreductases, laccases, lipases, P450 monooxygenases, and alkane hydroxylases [23]) have been proposed, which include conversion of the polymer to carboxylic acids and their further metabolism by $\beta$-oxidation and Krebs cycle [24]. The first proposed pathway is that styrene is converted into styrene epoxide by styrene monooxygenase and further into 4-maleylacetoacetate by styrene oxide isomerase, phenylacetaldehyde dehydrogenase, phenylacetate hydroxylase, 2-hydroxyphenylacetate hydroxylase, and homogentisate 1,2-dioxygenase. Through the $\beta$-oxidation pathway, 4-maleylacetoacetate is converted to acetyl-CoA, followed by citric acid (TCA) cycle to the central biosynthesis pathways. On the other hand, in the second proposed pathway of biodegradation of PS, styrene is hydroxylated by styrene dioxygenase on the aromatic ring to generate styrene cis-glycol, which can be further converted to acetyl-CoA by cis-glycol dehydrogenase, catechol 2,3-dioxygenase, 2-hydroxymuconic acid semialdehyde hydrolase, 2-hydroxypenta-2,4-dienoate hydratase, 4-hydroxy-2-oxovalerate aldolase, and pyruvate dehydrogenase complex. Similarly, acetyl-CoA will then enter the TCA cycle and participate in biomass synthesis or accumulation of other metabolites [17,25]. Another proposed pathway of biodegradation of PS starts from the side-chain cleavage of PS. Monooxygenases or aromatic ring hydroxylases are potential candidates to break the aromatic ring of PS. However, the detailed degradation pathway and involved enzymes have not still been revealed [17].

The most used bacteria for PS degradation are genera *Bacillus* [5,14,26,27], *Pseudomonas* [26], *Paenibacillus* [28], and *Rhodococcus* [29,30], while *Aspergillus* [5] and *Curvularia* [31] are the most investigated genera of fungi. Asmita et al. [5] reported PS weight loss of 20.0 and 5.0% by *Bacillus subtilis* and *Pseudomonas aeruginosa*, respectively, which were cultured in nutrient broth. In the other case, bacteria were cultured in Bushnell–Hass broth, and the weight loss of PS was higher (58.8%) for *Bacillus cereus*, but for *Pseudomonas aeruginosa,* weight loss was not observed. This indicates that providing a suitable nutrient medium for bacteria is important to enhance the biodegradation process. Auta et al. [14] investigated the biodegradation of PS microplastics by *Bacillus cereus* for 40 days. They reported a weight loss of PS of 7.4%. Moreover, a slightly higher weight loss was observed by Mohan et al. [26]. They investigated HIPS film degradation using *Bacillus* spp. and *Pseudomonas* sp. The weight loss percentage of HIPS film obtained after treatment with *Bacillus* spp. was significant at 23.7% and was less than 10.0% with *Pseudomonas* sp. Moreover, Kumar et al. [32] reported a PS weight loss of 34.0% by *Bacillus paralicheniformis* G1 after 60 days of exposure. However, the exposure of PS to *Rhodococcus ruber* for up to 8 weeks resulted in a small reduction in PS weight loss (0.8%). Mor and Sivan [30] demonstrated the high affinity of *Rhodococcus ruber* to PS, which led to biofilm formation and, presumably, induced partial biodegradation. Furthermore, *Pseudomonas* sp. was tested for PS degradation, and according to Kim et al. [33], colonies of *Pseudomonas* sp. were observed on the PS film after 30 days. Scanning electron microscopy analysis (SEM) revealed smoother edges and holes in the PS film. The viability and proliferation of *Pseudomonas* sp. DSM 50071 on the surface PS suggests that PS can be used as an energy resource and cellular component when no other alternative carbon sources are available. In addition, the study by Motta et al. [31] tested pretreated PS films. In order to induce changes in the structure of PS and thus facilitate the mechanism by which microorganisms can take up the carbon contained in the polymer, PS was first subjected to novel chemical oxidation treatments that can convert the polymer chains into more oxidized compounds with presumably lower molecular weights. These treatments trigger a series of physical and chemical changes that lead to the formation of carbonyl and hydroxyl groups. In one treatment, the oxidant was used alone (sample 1); in another, a transition metal complex was added to the oxidant (sample 2); and in the third, an inorganic acid was added to the oxidant (sample 3). In all treatments, the polymers were exposed to the oxidant for 2 h at room temperature and sterilized with UV radiation before incubation with the fungi. After pretreatment, the films were placed on plates containing Sabouraud agar on which the fungus *Curvularia* sp. was growing. After 4 weeks of exposure, *Curvularia* sp. began to colonize the surface of the film pretreated with the oxidizing agent. However, after a longer period of time (9 weeks), *Curvularia* sp. completely colonized the surface of the pretreated PS. Colonization of the non-pretreated PS did not occur. These results suggest that improvement in biodegradation is possible by pretreating the MP as well as by longer exposure time. Furthermore, acceleration of biodegradation can be achieved by co-metabolism. In this process, primary substrates (e.g., glucose) are added to the polluted medium to stimulate the microorganisms to produce degrading enzymes. However, this does not always have a positive effect on decomposition. For example, Shabbir et al. [34] reported that the addition of glucose increased the biodegradation of MP by the periphyton biofilm for all MP (from 9.52–18.02%, 5.95–14.02%, and 13.24–19.72% for PP, PE, and PET) after 60 days, respectively, while a mixture of peptone and glucose and peptone together had an inhibitory effect. Accordingly, there are a variety of factors that influence the biodegradation of MP. The study of key biodegradation factors can be achieved by the Taguchi experimental design, which allows a minimal number of experiments, saving time and resources. These are arrays selected according to the number of factors and levels. The influencing factors are arranged in orthogonal arrays, and the levels on which they are examined vary. The Taguchi method is most suitable for studying processes where there are few interactions between factors; some factors are statistically significant, and there is an average number of factors, such as 3 to 50 [35]. However, the study of optimal conditions is usually conducted by changing one variable while all other variables remain fixed in a given set of conditions. To find the correct optimum, possible interactions between variables should be considered. On the other hand, factorial experimental design can be used to estimate the interactions of possible influencing parameters on efficiency with a limited number of planned trials [36]. In addition, factorial experimental designs allow the investigation of the joint effect of factors on a response. Full factorial experiments consist of all possible combinations of values for all factors [37]. This experimental design is also commonly used in bioremediation studies [36,38].

This study investigated the significant factors affecting the biodegradation of MP-PS particles by bacteria *Bacillus cereus* and *Pseudomonas alcaligenes*. Experiments were conducted using the Taguchi method, and seven factors were investigated at two levels. After 30 days of experiments, the statistically significant factors for biodegradation were determined using colony-forming units (CFU) as the response parameter. Subsequently, biodegradation experiments were conducted with the obtained significant factors using the full factorial design to determine the optimal conditions for the biodegradation of PS by the two mentioned bacteria.

## 2. Materials and Methods

### 2.1. Microplastics Preparation

MP were obtained by grinding plastic disposable accessories such as spoons for PS. First, these materials were cut into smaller pieces with scissors and then ground in a cryo-mill (Retsch, Haan, Germany) with liquid nitrogen and dried in the air for 48 h at room temperature. Next, obtained particles were sieved on stainless steel screens (RX-86-1 Sieve shaker, W.S. Tyler, Mentor, OH, USA) to obtain particles in size ranges: 500–700 μm, 300–500 μm, and 100–300 μm. After sieving, the MP particles were stored in glass bottles. Before the experiments, MP particles were sterilized in 100 mL flasks containing 70% ethanol for 10 min on a rotary shaker (Heidolph unimax 1010, Heidolph incubator 1000, Schwabach, Germany) at 160 rpm. The particles were separated from the ethanol by vacuum membrane filtration (through cellulose nitrate 0.45 μm sterile filter (ReliaDisc$^{TM}$, Ahlstrom-Munksjö, Helsinki, Finland)) and additionally washed with sterile deionized water.

### 2.2. Bacterial Cultivation

The bacteria *Bacillus cereus* and *Pseudomonas alcaligenes* were isolated from environmental samples in which MP particles were present, from activated sludge (municipal wastewater treatment plant, Vrgorac, Croatia) and river sediment (Kupa, Croatia). In total, 100 mL of activated sludge and river sediment were placed in 250 mL sterilized Erlenmeyer flasks and shaken for 24 h at room temperature on the thermostatic rotary shaker at 160 rpm. After 24 h, the colony-forming units (CFU) of bacteria on the general-purpose medium (nutrient agar (NA)) were determined using the pour plate method according to Briški et al. [39]. For plate counting, a dilution series (0.9% mass aqueous NaCl solution) was prepared from each sample. The plates were incubated at 80% relative humidity and 37 °C to culture the bacteria. After incubation, the number of colonies on the agar plates was determined. Bacterial colonies that were morphologically distinct and dominated on the nutrient agar plates were collected and transferred to the nutrient agar plates and incubated at 37 °C for 24–48 h. After pure isolates were obtained, they were stored in slants for characterization. The isolated bacteria were Gram-stained (Figure 1) [40]. After Gram staining, a series of biochemical tests known as API (Analytical profile index, bioMérieux®, Lyon, France) were performed. Gram-negative bacteria were identified using API strip 20 E (Analytical Profile Index, bioMérieux®, Lyon, France). The final step of bacterial identification was matrix-assisted laser desorption/ionization time-of-flight mass spectrometry (Microflex LT MALDI-TOF MS, Bruker Daltonics, Bremen, Germany), which is based on protein identification of pulsed single ionic analytes (pure microbial culture) coupled with a TOF measurement mass analyzer, and the exact protein mass was determined. Cultures were set on the pre-cultivation in the mineral media [41] for 24 h before experiments in order to achieve the growth's log phase. The optical density (*OD*) of bacterial suspension was measured on a spectrophotometer (DR/2400 Portable Spectrophotometer, Hach, Loveland, CO, USA) at $\lambda$ = 600 nm, and CFU was determined by the decimal plate method [42].

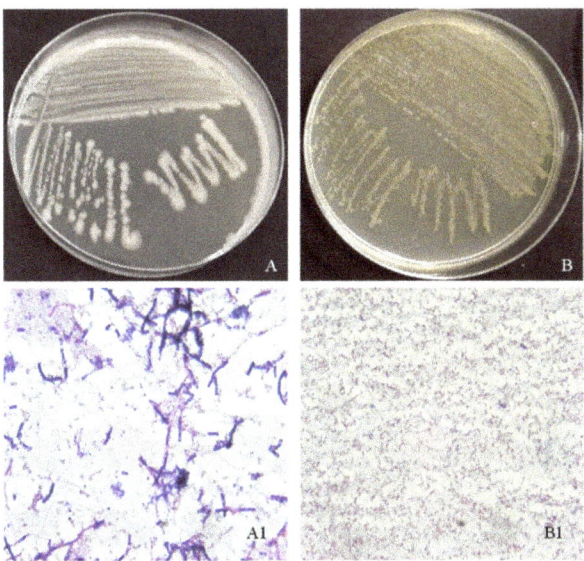

**Figure 1.** Bacteria (**A**) *Bacillus cereus* and (**B**) *Pseudomonas alcaligenes* cultivate on nutrient agar by streak plate method. Microphotos of bacterial Gram-stained smears, (**A1**) *Bacillus cereus*, and (**B1**) *Pseudomonas alcaligenes* at the microscope magnification of 1000×.

## 2.3. Biodegradation Experiments

Biodegradation of MP-PS particles was investigated by the Gram-positive bacterium *Bacillus cereus* and the Gram-negative bacterium *Pseudomonas alcaligenes*, respectively. The mentioned microorganisms were isolated from environmental samples where MP particles were found and adapted to the above conditions. Therefore, it can be assumed that the bioaugmentation of the autochthonous microorganisms into the system will accelerate the biodegradation process. The research was divided into two parts: preliminary experiments according to the Taguchi design (P1 and P2) and the main experiments by full factorial design (M1 and M2). All experiments were carried out in duplicate.

The experiments (preliminary and main) were conducted in 250 mL Erlenmeyer flasks with a working volume of 100 mL for 30 days at the thermostatic rotary shakers (Heidolph unimax 1010, Heidolph incubator 1000, Schwabach, Germany). The temperatures and agitation speeds on the rotary shakers were set according to the experimental plan. The flasks contained mineral medium (composition according to [41]), bacterial suspension (*Bacillus cereus* or *Pseudomonas alcaligenes*), and MP-PS. The control flasks were set for the purpose of monitoring bacterial growth and contain mineral medium and bacterial suspension without MP-PS particles.

### 2.3.1. Preliminary Experiments

Preliminary experiments were performed according to the $L_8$ orthogonal array listed in Table 1 and obtained by the Taguchi design (7 factors at 2 levels). Significant factors (pH-value, temperature, $T$, MP size, concentration of MP, $\gamma_{MP}$, agitation speed, optical density of bacterial suspension, $OD$, and addition of glucose, $\gamma_{GLU}$) that can influence the biodegradation process of MP-PS by *Bacillus cereus* and *Pseudomonas alcaligenes* were investigated. Experiments (presented in Table 1) were marked as P1 (1-1 to 1-8) and P2 (2-1 to 2-8) for MPs-PS biodegradation by *Bacillus cereus* and *Pseudomonas alcaligenes*, respectively. In experiments with *Bacillus cereus*, CFU were $5.5 \times 10^6$ cells/mL and $2.6 \times 10^7$ cells/mL at optical densities of 0.1 and 0.5, respectively. In experiments with *Pseudomonas alcaligenes*,

CFU values were $4.5 \times 10^7$ cells/mL and $3.5 \times 10^8$ cells/mL at optical densities of 0.1 and 0.5, respectively.

**Table 1.** Taguchi $L_8$ orthogonal array for each factor at 2 levels.

| Experiment No. | pH/- | $T$/°C | MP Size/μm | $\gamma_{MP}$/mg/L | Agitation Speed/rpm | $OD$/- | $\gamma_{GLU}$/mg/L |
|---|---|---|---|---|---|---|---|
| P1-1; P2-1 | 8 | 15 | 600 | 1000 | 100 | 0.5 | 0 |
| P1-2; P2-2 | 6 | 15 | 200 | 50 | 100 | 0.1 | 0 |
| P1-3; P2-3 | 6 | 15 | 200 | 1000 | 200 | 0.5 | 100 |
| P1-4; P2-4 | 8 | 15 | 600 | 50 | 200 | 0.1 | 100 |
| P1-5; P2-5 | 6 | 25 | 600 | 50 | 100 | 0.5 | 100 |
| P1-6; P2-6 | 8 | 25 | 200 | 1000 | 100 | 0.1 | 100 |
| P1-7; P2-7 | 6 | 25 | 600 | 1000 | 200 | 0.1 | 0 |
| P1-8; P2-8 | 8 | 25 | 200 | 50 | 200 | 0.5 | 0 |

Analyses of key process parameters, by contribution, for PS biodegradation by *Bacillus cereus* and *Pseudomonas alcaligenes* were determinate using the logarithmic CFU value (log CFU) as a response parameter. The influence of each of the above factors was assessed using the $L_8$ orthogonal array method, and the factors with the greatest contribution by analysis of variance were determined (ANOVA). Statistical significance of the factors' effects was considered at a 95.0% confidence level. Calculations and statistical analyses were performed using Design-Expert 10.0, Stat-Ease, Minneapolis, MN, USA.

2.3.2. Main Experiments

Preliminary experiments pointed out three factors with the highest contribution to biodegradation. Based on the results of the preliminary experiments, the authors designed two new sets of experiments (M1 and M2 for *Bacillus cereus* and *Pseudomonas alcaligenes*, respectively) to determine the optimal conditions for MP-PS biodegradation. The experiments were designated following the full factorial methodology with 3 factors at 3 levels, resulting in a total of 27 experiments each (Table 2). M1 experiments included MP concentration ($\gamma_{MP}$), MP size, and agitation speed as factors. The pH-value of the mineral medium (7.323), temperature ($25 \pm 0.2$ °C), and optical density ($OD = 0.3$) were constant. In the case of M2, the factors were optical density, agitation speed, and size of MP particles; the pH-value of the mineral medium (7.323), temperature ($25 \pm 0.2$ °C), and concentration of MP (500 mg/L) were constant in this set of experiments. In all experiments (M1 and M2), glucose was not added to the media. Calculations and statistical analyses were performed using Design-Expert 10.0, Stat-Ease, Minneapolis, MN, USA.

**Table 2.** Full factorial experimental design for MP-PS biodegradation process by bacteria *Bacillus cereus* and *Pseudomonas alcaligenes* at 3 levels.

| | *Bacillus cereus* | | | *Pseudomonas alcaligenes* | | | |
|---|---|---|---|---|---|---|---|
| Experiment No. | $\gamma_{MP}$/mg/L | MP Size/μm | Agitation Speed/rpm | Experiment No. | Agitation Speed/rpm | MP Size/μm | $OD$/- |
| M1-1 | 1000 | 400 | 100 | M2-1 | 200 | 400 | 0.1 |
| M1-2 | 50 | 400 | 200 | M2-2 | 100 | 400 | 0.5 |
| M1-3 | 1000 | 200 | 150 | M2-3 | 200 | 200 | 0.3 |
| M1-4 | 50 | 200 | 100 | M2-4 | 100 | 200 | 0.1 |
| M1-5 | 500 | 400 | 100 | M2-5 | 150 | 400 | 0.1 |
| M1-6 | 50 | 600 | 150 | M2-6 | 100 | 600 | 0.3 |
| M1-7 | 50 | 600 | 100 | M2-7 | 100 | 600 | 0.1 |
| M1-8 | 500 | 200 | 100 | M2-8 | 150 | 200 | 0.1 |
| M1-9 | 500 | 600 | 100 | M2-9 | 150 | 600 | 0.1 |
| M1-10 | 1000 | 200 | 200 | M2-10 | 200 | 200 | 0.5 |
| M1-11 | 1000 | 200 | 100 | M2-11 | 200 | 200 | 0.1 |

**Table 2.** *Cont.*

| | *Bacillus cereus* | | | | *Pseudomonas alcaligenes* | | |
|---|---|---|---|---|---|---|---|
| Experiment No. | $\gamma_{MP}$/mg/L | MP Size/µm | Agitation Speed/rpm | Experiment No. | Agitation Speed/rpm | MP Size/µm | OD/- |
| M1-12 | 50 | 200 | 150 | M2-12 | 100 | 200 | 0.3 |
| M1-13 | 500 | 400 | 200 | M2-13 | 150 | 400 | 0.5 |
| M1-14 | 500 | 400 | 150 | M2-14 | 150 | 400 | 0.3 |
| M1-15 | 1000 | 600 | 150 | M2-15 | 200 | 600 | 0.3 |
| M1-16 | 50 | 600 | 200 | M2-16 | 100 | 600 | 0.5 |
| M1-17 | 50 | 200 | 200 | M2-17 | 100 | 200 | 0.5 |
| M1-18 | 1000 | 600 | 200 | M2-18 | 200 | 600 | 0.5 |
| M1-19 | 1000 | 600 | 100 | M2-19 | 200 | 600 | 0.1 |
| M1-20 | 500 | 600 | 150 | M2-20 | 150 | 600 | 0.3 |
| M1-21 | 500 | 600 | 200 | M2-21 | 150 | 600 | 0.5 |
| M1-22 | 1000 | 400 | 200 | M2-22 | 200 | 400 | 0.5 |
| M1-23 | 1000 | 400 | 150 | M2-23 | 200 | 400 | 0.3 |
| M1-24 | 50 | 400 | 100 | M2-24 | 100 | 400 | 0.1 |
| M1-25 | 50 | 400 | 150 | M2-25 | 100 | 400 | 0.3 |
| M1-26 | 500 | 200 | 150 | M2-26 | 150 | 200 | 0.3 |
| M1-27 | 500 | 200 | 200 | M2-27 | 150 | 200 | 0.5 |

During these experiments, *OD* and *CFU* were measured in order to analyze bacterial growth. After 30 days of the biodegradation process, the optimal conditions for MP-PS biodegradation by *Bacillus cereus* and *Pseudomonas alcaligenes* were determinate using the log CFU as a response parameter.

Additional analyses of the aqueous phase during the experiments were also performed to verify the biodegradation process and to get a better insight into the whole process. Total carbon (TC), total organic carbon (TOC), and inorganic carbon (IC) were determined (TOC-VCSH, Shimadzu, Japan). Released additives in the aqueous phase were estimated by LC/MS analysis (LC-MS 2020, Shimadzu, Japan).

At the end of the experiments, the MP-PS particles were separated from the aqueous phase by vacuum membrane filtration (and washed with sterile deionized water). The ecotoxicity of the aqueous phase (Lumistox 300, Dr. Lange GmbH, Düsseldorf, Germany) with the marine bacterium *Vibrio fischeri* [43] was conducted.

MP-PS particles were analyzed before and after biodegradation by FTIR-ATR spectroscopy (Spectrum One, PerkinElmer, Waltham, MA, USA).

*2.4. Response Surface Modeling*

Response surface modeling (RSM) was applied to define the influence of the concentration of MP particles ($X_1$), particles size ($X_2$), and agitation speed ($X_3$) on the logarithmic number of the living cells (log CFU) of *Bacillus cereus* in M1 experiments. For M2 experiments, it was necessary to define the influence of agitation speed ($X_1$), particles size ($X_2$), and OD ($X_3$) on the logarithmic number of the living cells (log CFU) of *Pseudomonas alcaligenes*. For that purpose, the MP size intervals were replaced by corresponding average values: 200, 400, and 600 µm. Two polynomials of various complexities were applied to describe the response surface. The models were presented by Equations (1) and (2).

$$\log CFU = a_0 + a_1 X_1 + a_2 X_2 + a_3 X_3 \tag{1}$$

$$\log CFU = a_0 + a_1 X_1 + a_2 X_2 + a_3 X_3 + a_4 X_1 \cdot X_2 + a_5 X_1 \cdot X_3 + a_6 X_2 \cdot X_3 + a_7 X_1{}^2 + a_8 X_2{}^2 + a_9 X_3{}^2 \tag{2}$$

Letter *a*, used in these models, represents model coefficients. MODEL I (Equation (1)) contains linear contributions of the concentration/agitation speed, particle size, and agitation speed/optical density for M1 and M2 experiments, respectively. MODEL II (Equation (2)) is, in fact, MODEL I upgraded by the interaction terms ($X_1 \cdot X_2$, $X_1 \cdot X_3$, and

$X_2 \cdot X_3$) as well as by quadratic terms ($X_1{}^2$, $X_2{}^2$, and $X_3{}^2$). Calculations and analyses were performed using Design-Expert 10.0, Stat-Ease, Minneapolis, MN, USA.

## 3. Results and Discussion

### 3.1. Preliminary Experiment

In the preliminary experiments, seven factors ($T$, pH-value, $\gamma_{MP}$, $\gamma_{GLU}$, agitation speed, size of MP, and $OD$) were examined on two levels according to the Taguchi design, Table 1. According to the response parameter, log CFU (Table 3), the statistically significant factors for the biodegradation of MP-PS by *Bacillus cereus* were the size of MP (48.52%), $\gamma_{MP}$ (5.03%), and agitation speed (41.31%). From the results according to the contribution, the agitation speed and the size of MP-PS had the greatest influence on the biodegradation of MP-PS by *Bacillus cereus*. This is not surprising, as the smaller particle size and larger surface area contribute to the bacterial colonization of the MP-PS particles, and colonization is considered the first necessary step for biodegradation [25]. In addition, the rotary shaker's agitation speed maintains the dissolved oxygen concentration, which is essential for biochemical reactions [44]. Furthermore, increased oxygen concentration accelerates the degradation of the polymer [45]. According to the oxygen demand, *Bacillus cereus* is a facultative anaerobe that can adapt to and grow in anoxic conditions [46]. Moreover, the presence of higher concentrations of MP-PS can negatively affect the biodegradation process due to its toxic effects, such as reducing the efficiency of the photosynthesis process and damaging the cells of microorganisms [47]. Over time, they can release various additives (plasticizers, stabilizers, pigments, fillers, and flame retardants) that have been shown to have a toxic effect on organisms [48,49]. In addition, this rod-shaped bacterium forms endospores that make it more resistant to extreme environmental conditions and enable its growth, adaptability, and survival [46].

**Table 3.** Results of preliminary experiments designed by the Taguchi method for biodegradation of MP-PS by *Bacillus cereus* and *Pseudomonas alcaligenes*.

| | *Bacillus cereus* | | | *Pseudomonas alcaligenes* | | |
|---|---|---|---|---|---|---|
| Factors | $\gamma_{MP}$ | MP Size | Agitation Speed | Agitation Speed | MP Size | $OD$ |
| Sum of Squares | 0.07 | 0.68 | 0.58 | 1.88 | 0.51 | 0.49 |
| $DF$ * | 1 | 1 | 1 | 1 | 1 | 1 |
| $F$-value | 128.82 | 1243.26 | 1058.59 | 46.93 | 12.72 | 12.22 |
| $p$-value | 0.002 | 0.000 | 0.000 | 0.006 | 0.038 | 0.040 |
| Contribution/% | 5.03 | 48.52 | 41.31 | 14.48 | 14.48 | 13.91 |

\* $DF$ = degree of freedom.

According to Table 3, significant factors for the biodegradation of MP-PS by *Pseudomonas alcaligenes* were agitation speed (53.42%), size of MP-PS (14.48%), and $OD$ (13.91%). However, the agitation speed and size of MP-PS had a higher impact on the biodegradation process due to higher percentages of contribution. As mentioned above, the agitation speed is important to ensure the dissolved oxygen concentration in the system. Apart from this effect, it also allows homogenization and bioavailability of particles to bacteria. Indeed, *Pseudomonas alcaligenes* is an aerobic bacterium, which means that the dissolved oxygen concentration is essential for its growth [50]. This confirms our results, as the better biodegradation of PS by *Pseudomonas alcaligenes* occurred at the higher tested agitation speed (200 rpm). *Pseudomonas alcaligenes* is a Gram-negative, rod-shaped bacterium commonly used in bioremediation for the degradation of polycyclic aromatic hydrocarbons [51]. One of the many factors that influence the biodegradation of MP-PS is particle size; a larger particle size means a smaller surface area, which directly reduces the possibility of colonization (biofilm formation) of MP-PS by bacteria [25]. In addition, $OD$ represents the number of live and dead bacterial cells and is a useful and technically simple parameter to indicate bacterial growth. A higher $OD$ value means a higher number of bacterial cells, which consequently increases the efficiency of the biodegradation process. However, CFU determination is a better indicator of bacterial growth.

### 3.2. Main Experiment

Optimal conditions for biodegradation of MP-PS were determined from the main experiments, M1 and M2, for *Bacillus cereus* and *Pseudomonas alcaligenes*, respectively.

To monitor bacterial growth, CFU was determined during the biodegradation of MP-PS (Figure 2). CFU as the number of live bacterial cells was increased until days 7 and 14 of the experiment M1-24 and M2-23, respectively. CFU of *Bacillus cereus* (experiment M1-24) increased from the initial value ($9.3 \times 10^6$ cells/mL) at the beginning of the experiment to $3.2 \times 10^7$ cells/mL on the 7th day. The same trend was observed in M2-23 for *Pseudomonas alcaligenes*, which showed a higher increase (from an initial CFU of $6.5 \times 10^7$ cells/mL to $2.3 \times 10^8$ cells/mL on day 14). After the exponential phase of bacterial growth, the stationary phase occurred (Figure 2A,B) in which the number of live and dead bacterial cells is equal. On the last day (day 30) of the experiments, the CFU was $2.2 \times 10^7$ cells/mL and $2.8 \times 10^8$ cells/mL for *Bacillus cereus* and *Pseudomonas alcaligenes*, respectively. Accordingly, the studied conditions were favorable for the growth and multiplication of the bacteria due to the possible production of degradation products. In addition, the CFU values of the control were lower compared to the samples containing MP-PS. This suggests that bacteria have adapted to the conditions with MP-PS and are likely using MP-PS as a carbon source for growth. This ability of bacteria of the genera *Bacillus* and *Pseudomonas* has been investigated by other researchers [14,26,33].

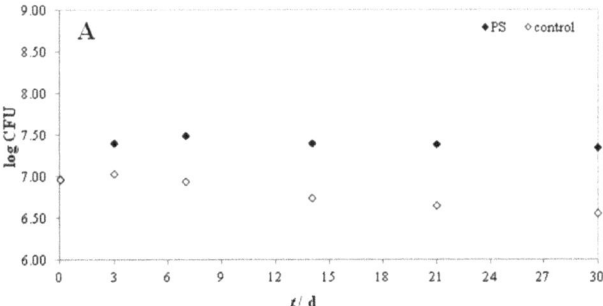

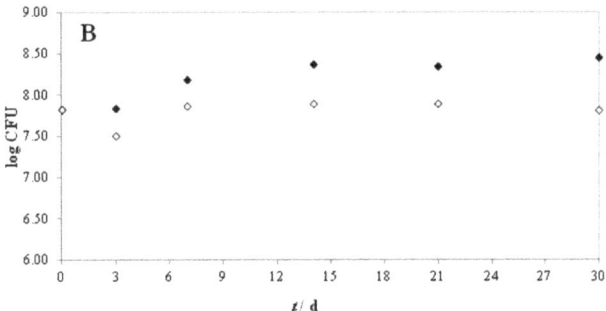

**Figure 2.** Changes of log CFU of (**A**) *Bacillus cereus* in experiment M1-24 and (**B**) *Pseudomonas alcaligenes* in experiment M2-23 during 30 days of MP-PS biodegradation.

The same trend as CFU was observed in the changes of TOC and TIC values of the aqueous phase in M1-24 (Figure 3A) and M2-23 (Figure 3B). The TOC and TIC values increased until the 7th day. After the 7th day and until the end of the experiment, the TOC and TIC values had not changed significantly. This is consistent with the growth stages of *Bacillus cereus* shown in Figure 2A. In experiment M2-23, TOC and TIC values increased until day 14. These changes were also consistent with the changes in CFU of *Pseudomonas alcaligenes* during the 30-day biodegradation of PS. However, TOC and TIC concentrations in the blank (BP) also increased, indicating the lysis of bacterial cells [27,52]. Compared to BP, TOC and TIC values were higher in samples with MP-PS particles, indicating the utilization of MP-PS by bacteria. Higher TOC and TIC values were observed for samples with *Pseudomonas alcaligenes*, correlating with higher CFU values for this bacterium. This indicates more efficient biodegradation of MP-PS by *Pseudomonas alcaligenes* than *Bacillus cereus*. An increase in TOC concentration in the sample may indicate the production of degradation products and/or the release of additives from the surface MP-PS. The increase in TOC concentration in the sample correlates with an increase in TIC levels, indicating that biodegradation has occurred [52]. In general, the TOC content of polluted water decreases during biodegradation [53]. This trend usually correlates with the increase in TIC since $CO_2$ is a product of mineralization [54]. However, during the biodegradation of water containing solid plastic particles, the TOC content may not decrease due to the plastic particles' release of additives or/and synthetic polymer analogs. These compounds may be mineralized to $CO_2$, which is reflected in an increase in the TIC value. This is consistent with the LC/MS analysis results of the aqueous phases (Figure 4).

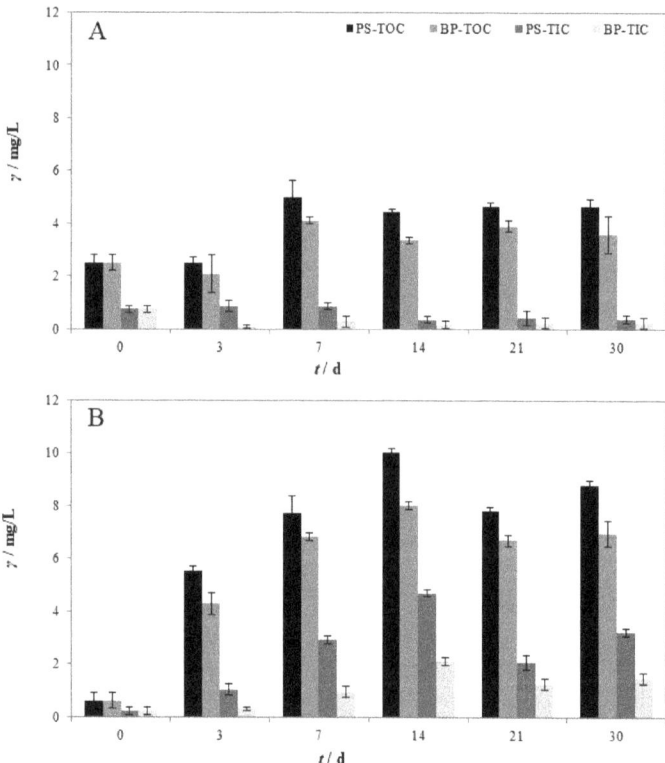

**Figure 3.** Changes in concentration ($\gamma$) of total organic carbon (TOC) and total inorganic carbon (TIC) for blank (BP) and sample (MP-PS) during 30 days of exposure to: (**A**) *Bacillus cereus* in experiment M1-24 and (**B**) *Pseudomonas alcaligenes* in experiment M2-23.

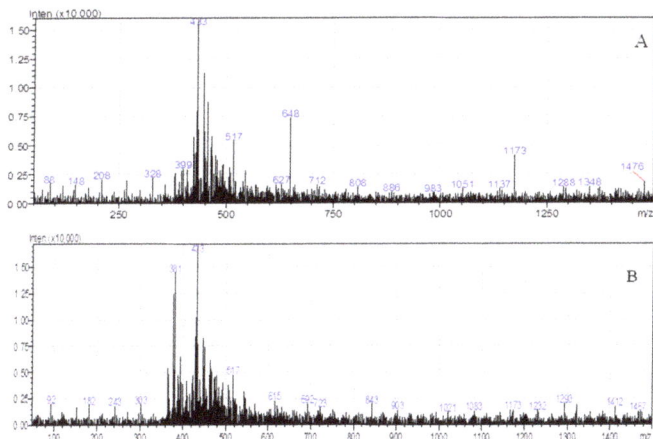

**Figure 4.** LC/MS analysis of aqueous phase for biodegradation process of MP-PS by (**A**) *Bacillus cereus* in experiment M1-24 and (**B**) *Pseudomonas alcaligenes* in experiment M2-23.

During the exposure of *Bacillus cereus* and *Pseudomonas alcaligenes* to MP-PS particles, additives that were added to the material during production may be released in the aqueous phase, and that was monitored by a flow injection LC/MS analysis. At the beginning of the experiment, no peaks were recorded on the chromatogram. However, on the 7th day, peaks appeared in the chromatogram (Figure 4), indicating the release of the additive from the surface MP. LC/MS analyses of the aqueous phase for the M1-24 experiment (Figure 4A) indicate the presence of [M+H]+ product at $m/z$ peak of 328 that may represent triphenyl phosphate, flame retardant in plastics [55]. Triphenyl phosphate is a commonly used commercial chemical additive that is classified as an organophosphate flame retardant and poses a potential toxic risk to aquatic organisms [56]. In the sample of M2-23, an antioxidant additive butylated hydroxytoluene was assumed at an $m/z$ ratio of 243 (Figure 4B) in the form [M+Na]+ [55]. Butylated hydroxytoluene, as a phenolic compound, is one of the frequent antioxidants used to protect plastics against oxidation (during their exposure to heat and light) [57]. Generally, according to Ho et al. [4] and De-la-Torre et al. [58], various additives such as antioxidants, UV stabilizers, processing lubes, antistats, and flame retardants may be incorporated into MP-PS. These compounds improve MP-PS properties but cause serious ecotoxicological concerns for the water environment.

The ecotoxicity test was performed with *Vibrio fischeri* for the aqueous phases of experiments M1-24 and M2-23, which are shown in Table 4. The inhibition value (*INH*) of the aqueous phase for M2-23 was higher than the *INH* value for M1-24, indicating higher toxicity of the sample obtained after the biodegradation of PS by *Pseudomonas alcaligenes*. The lower $EC_{20}$ value in M1-24 also indicates higher toxicity of the mentioned sample. These results confirm the previously mentioned findings regarding the higher CFU, TOC, and TIC values in M2-23 than in M1-24 and the released additives. The aqueous phase of M2-23 contained some degradation products and/or additives that may be toxic to aquatic organisms. However, these obtained toxicity values are relatively low due to the not possible estimation of $EC_{50}$.

**Table 4.** Inhibition of aqueous phase (*INH*) and *EC* values ($EC_{20}$) obtained by ecotoxicity tests for the aqueous phase in experiments M1-24 and M2-23 by *Vibrio fischeri*.

| Experiment No. | INH/% | $EC_{20}$/% |
|---|---|---|
| M1-24 | 40.11 | 28.57 |
| M2-23 | 46.79 | 27.40 |

For the purpose of confirming MP-PS biodegradation, FTIR-ATR analyses were carried out. Characteristic MP-PS peaks at 3024, 2847, 1601, 1492, 1451, 1027, and 694 $cm^{-1}$ [59] are in Figure 5. The peak of 3024 $cm^{-1}$ is specific for aromatic C−H stretching while stretching of the other C−H groups is detected at 2847 $cm^{-1}$. Wavenumbers 1601 and 1492 $cm^{-1}$ are related to aromatic ring stretching. The bending of the $CH_2$ group occurs at 1451 $cm^{-1}$, while the bending of the aromatic C−H groups has characteristic peaks at 1027 and 694 $cm^{-1}$ [59]. All characteristic peaks decreased their intensities after biodegradation by *Bacillus cereus* (M1-24) (Figure 5A). The decrement in intensities of characteristic FTIR-ATR peaks was noticed after treatment by *Pseudomonas alcaligenes* (M2-23), as well (Figure 5B). The peak between 1000 to 750 $cm^{-1}$, representing C−H groups, disappeared in both spectra, while the new peak at approximately 1395 $cm^{-1}$ appeared; Subramani and Sepperumal [60] noted the appearance of the same peak during the biodegradation of PS foam by *Pseudomonas* sp. During the oxidation process, functional groups such as hydroxyl or carbonyl groups could be formed via β-oxidation, which are known to be used in the TCA cycle or in the energy metabolism of bacteria, thereby increasing hydrophilicity [61,62]. A characteristic weak peak at 1715 $cm^{-1}$ was detected in both spectra, indicating the formation of carbonyl groups (−C=O). Therefore, the increased number of oxygen atoms on the plastic surface in areas exhibiting bacterial growth is direct evidence of PS degradation [16].

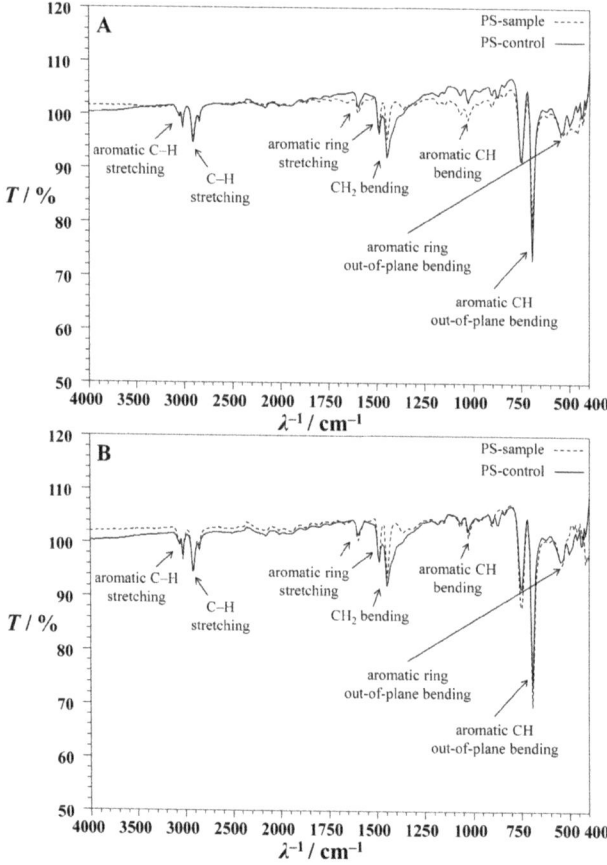

**Figure 5.** FTIR-ATR spectroscopy of MP-PS particles before (MP-PS control) and after (MP-PS sample) biodegradation by: (**A**) *Bacillus cereus* in experiment M1-24 and (**B**) *Pseudomonas alcaligenes* in experiment M2-23.

According to significant parameters (Table 3), the optimal conditions for MP-PS biodegradation were investigated by full factorial design. Three factors were examined at three levels (Table 5), and response surfaces (Figure 6) were created to determine the effect of the factors on changes in log CFU. As can be seen in Figure 6A, the concentration and size of MP-PS have no significant effect on the growth of bacteria (CFU). In contrast, the living colonies of *Bacillus cereus* (CFU) increased with decreasing agitation speed and MP-PS concentration (red parts of response surfaces in Figure 6B,C). These results indicate a greater influence of agitation speed on the biodegradation of MP-PS by *Bacillus cereus*. In the case of biodegradation of MP-PS by *Pseudomonas alcaligenes*, the number of bacteria colonies (CFU) increased with increasing agitation speed and at a particle size of PS = 400 μm (Figure 6D). Figure 6E shows an increase in the number of live bacteria colonies with increasing agitation speed and $OD = 0.3$. According to the red area in Figure 6F, the highest log CFU was observed at $OD = 0.3$, and the mean investigated particle size of MP-PS (200 μm). This indicates a high effect of all three investigated factors on CFU values, which was confirmed by ANOVA analysis (Table 5).

**Table 5.** ANOVA analysis for obtained models for MP-PS biodegradation by *Bacillus cereus* and *Pseudomonas alcaligenes*.

| Bacterium | Applied Model | Statistical Analysis | | | | | | Influential Model Factors | Influential Parameters |
|---|---|---|---|---|---|---|---|---|---|
| | | Model | | | | Coefficients | | | |
| | | $R^2$ | $R^2_{adj}$ | $F$ | $p$ | Coefficient Value | $p$ | | |
| *Bacillus cereus* | MODEL I | 0.8616 | 0.8435 | 47.71 | 0.000 | $a_0 = 7.55$ | 0.000 | | agitation speed |
| | | | | | | $a_1 = 1.34 \times 10^{-4}$ | 0.304 | | |
| | | | | | | $a_2 = -9.02 \times 10^{-5}$ | 0.107 | | |
| | | | | | | $a_3 = -6.03 \times 10^{-3}$ | 0.000 | $X_3$ | |
| | MODEL II | 0.9108 | 0.8636 | 19.29 | 0.000 | $a_0 = 8.16$ | 0.000 | | agitation speed |
| | | | | | | $a_1 = 1.30 \times 10^{-3}$ | 0.271 | | |
| | | | | | | $a_2 = -1.57 \times 10^{-5}$ | 0.091 | | |
| | | | | | | $a_3 = -0 \times 02$ | 0.000 | $X_3$ | |
| | | | | | | $a_4 = 1.87 \times 10^{-7}$ | 0.550 | | |
| | | | | | | $a_5 = -1.02 \times 10^{-6}$ | 0.732 | | |
| | | | | | | $a_6 = -8.38 \times 10^{-7}$ | 0.504 | | |
| | | | | | | $a_7 = -1.38 \times 10^{-6}$ | 0.198 | | |
| | | | | | | $a_8 = -2.23 \times 10^{-8}$ | 0.905 | | |
| | | | | | | $a_9 = 4.25 \times 10^{-5}$ | 0.020 | $X_3^2$ | |
| *Pseudomonas alcaligenes* | MODEL I | 0.4287 | 0.3542 | 5.75 | 0.004 | $a_0 = 7.44$ | 0.000 | | agitation speed and OD |
| | | | | | | $a_1 = 3.87 \times 10^{-3}$ | 0.011 | $X_1$ | |
| | | | | | | $a_2 = -3.67 \times 10^{-4}$ | 0.304 | | |
| | | | | | | $a_3 = 1.10$ | 0.008 | $X_3$ | |
| | MODEL II | 0.9076 | 0.8587 | 18.55 | 0.000 | $a_0 = 3.49$ | 0.000 | | agitation speed, MP size and OD |
| | | | | | | $a_1 = 4.98 \times 10^{-2}$ | 0.000 | $X_1$ | |
| | | | | | | $a_2 = 4.37 \times 10^{-4}$ | 0.038 | $X_2$ | |
| | | | | | | $a_3 = 6.93$ | 0.000 | $X_3$ | |
| | | | | | | $a_4 = 2.25 \times 10^{-6}$ | 0.581 | | |
| | | | | | | $a_5 = -1.04 \times 10^{-2}$ | 0.018 | $X_1 \cdot X_3$ | |
| | | | | | | $a_6 = 1.60 \times 10^{-3}$ | 0.127 | | |
| | | | | | | $a_7 = -1.46 \times 10^{-4}$ | 0.000 | $X_1^2$ | |
| | | | | | | $a_8 = -2.03 \times 10^{-6}$ | 0.169 | | |
| | | | | | | $a_9 = -8.32$ | 0.000 | $X_3^2$ | |

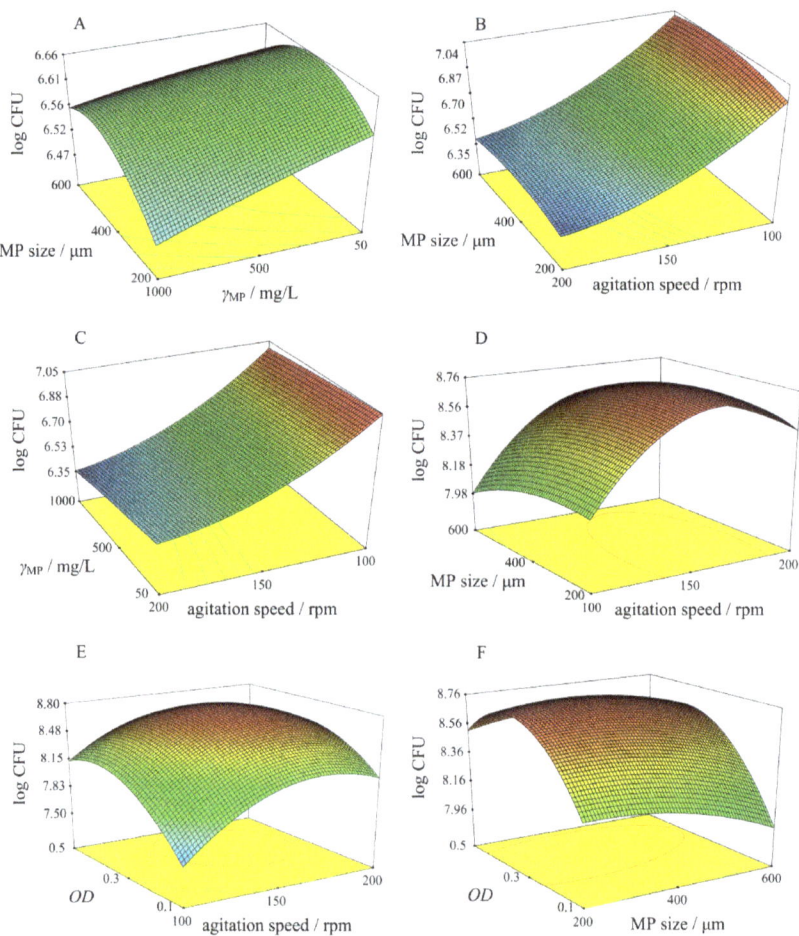

**Figure 6.** Response surfaces obtained for MP-PS biodegradation by *Bacillus cereus* (**A–C**) and by *Pseudomonas alcaligenes* (**D–F**) in main experiments by full factorial design.

The analysis of the developed mathematical models revealed information about the biodegradation of MP-PS by *Bacillus cereus*, Table 5. Two models (linear and quadratic) were used to describe the experimental data. MODEL I showed the agitation speed of the rotary shaker (term $X_3$) as the influential factor. The associated coefficient of determination had a relatively low value ($R^2 = 0.8616$) compared to more complex polynomials such as MODEL II. Following the highest value of $R^2$ for MODEL II, this model was the most statistically significant in describing the biodegradation of MP-PS particles. Statistical analysis of MODEL II pointed out that the agitation speed of the rotary shaker (terms $X_3$ and $X_3^2$) was the only factor influencing the MP-PS biodegradation. Moreover, statistical analysis of the biodegradation experiments with *Pseudomonas alcaligenes* shows that MODEL II proved to be the best model to describe this system (Table 5). The higher values of $R^2$ and $R^2_{adj}$ for MODEL II suggest that the statistical analysis is much closer to the experimental data, indicating the greater significance of MODEL II compared with MODEL I. In this case, the large, influential factors were agitation speed, size of MP-PS particles, *OD*, the interaction term of agitation speed and *OD*, as well as quadratic terms of agitation speed and *OD* (terms $X_1$, $X_2$, $X_3$, $X_1 \cdot X_3$, $X_1^2$, and $X_3^2$, respectively).

Overall, optimal biodegradation conditions are represented in Table 6. These results are in agreement with previously explained response surfaces and statistical analysis. Accordingly, the most efficient biodegradation of MP-PS by *Bacillus cereus* can be obtained at low MP-PS concentration and some average MP-PS particle size at 100 rpm. On the other hand, an average MP-PS particle size at a higher agitation speed and the average value of *OD* was the most suitable conditions for *Pseudomonas alcaligenes*. From the determined optimal conditions for biodegradation, the optimal particle size of MP is medium in both cases, which is surprising considering that the smaller particle size has a larger surface area. Thus, the bioavailability of the particles for biodegradation is better. However, according to previous studies, the toxic effect of MP increases with decreasing particle size [63,64].

**Table 6.** Optimal conditions for MP-PS biodegradation by *Bacillus cereus* and *Pseudomonas alcaligenes*.

| | *Bacillus cereus* | | | *Pseudomonas alcaligenes* | | |
|---|---|---|---|---|---|---|
| Factor | $\gamma_{MP}$/mg/L | MP Size/$\mu$m | Agitation Speed/rpm | Agitation Speed/rpm | MP Size/$\mu$m | *OD*/- |
| Value | 66.20 | 413.29 | 100.45 | 161.08 | 334.73 | 0.35 |

## 4. Conclusions

This research investigated the biodegradation of MP-PS particles by the bacteria *Bacillus cereus* and *Pseudomonas alcaligenes*. There are many factors that influence the biodegradation process of MP-PS. Hence, it is necessary to ensure optimal conditions for bacteria to stimulate bacteria for MP-PS biodegradation. Accordingly, the key factors that significantly influence these two bacteria's biodegradation of MP-PS were investigated. After this step, the determination of optimal conditions for MP-PS biodegradation was studied. In conclusion, the agitation speed of the rotary shaker plays a key role during MP-PS biodegradation for both bacteria. Furthermore, additional analyses provided a better understanding of the biodegradation process. Bacterial growth was monitored by determining the CFU; CFU changes were correlated with TOC values; an increase in TOC was observed due to the deterioration of PS structure and release of additives, which were correlated with LC/MS results. According to the $m/z$ peaks, the flame-retardant triphenyl phosphate and the antioxidant butylated hydroxyl toluene were noticed. The increase in TIC values indicates the formation of $CO_2$, which is a product of biodegradation. By FTIR-ATR spectroscopy, the deterioration of the MP-PS structure was confirmed due to the carbonyl group formation in both used bacteria. Overall, the obtained results indicate a higher biodegradation potential of MP-PS for *Pseudomonas alcaligenes*. However, *Bacillus cereus* and *Pseudomonas alcaligenes* are suitable choices for biodegrading MP-PS particles due to their great adaptability to various conditions. Future research should be focus on investigating these optimal biodegradation conditions.

**Author Contributions:** Conceptualization, Š.U., D.K.G. and T.B.; writing—original draft preparation, M.M. and M.C.; Writing—review and editing, D.K.G., T.B. and Š.U. All authors have read and agreed to the published version of the manuscript.

**Funding:** The authors would like to acknowledge the financial support of the Croatian Science Foundation through project Advanced Water Treatment Technologies for Microplastics Removal (AdWaTMiR) (IP-2019-04-9661).

**Institutional Review Board Statement:** Not applicable.

**Informed Consent Statement:** Not applicable.

**Data Availability Statement:** The data presented in this study are available on request from the corresponding author.

**Conflicts of Interest:** The authors declare no conflict of interest.

## References

1.  Plastics—The Facts 2021: An Analysis of European Plastics Production, Demand and Waste Data. Available online: https://plasticseurope.org/wp-content/uploads/2021/12/Plastics-the-Facts-2021-web-final.pdf (accessed on 4 January 2021).
2.  Bule, K.; Zadro, K.; Tolić, A.; Radin, E.; Miloloža, M.; Ocelić Bulatović, V.; Kučić Grgić, D. Mikroplastika u morskom okolišu Jadrana. *Chem. Ind.* **2020**, *69*, 303–310. [CrossRef]
3.  Tareen, A.; Seed, S.; Iqbal, A.; Batool, R.; Jamil, N. Biodeterioration of microplastics: A promising step towards plastics waste management. *Polymers* **2022**, *14*, 2275. [CrossRef] [PubMed]
4.  Ho, B.T.; Roberts, T.K.; Lucas, S. An overview on biodegradation of polystyrene and modified polystyrene: The microbial approach. *Crit. Rev. Biotechnol.* **2018**, *38*, 308–320. [CrossRef] [PubMed]
5.  Asmita, A.; Shubhamsingh, T.; Tejashree, S. Isolation of plastic degrading microorganisms from soil samples collected at various locations in Mumbai, India. *Int. Res. J. Environ. Sci.* **2015**, *4*, 77–85.
6.  Johnston, B.; Radecka, I.; Hill, D.; Chiellini, E.; Ivanova Ilieva, I.; Sikorska, W.; Musioł, M.; Zięba, M.; Marek, A.A.; Keddie, D.; et al. The microbial production of polyhydroxyalkanoates from waste polystyrene fragments attained using oxidative degradation. *Polymers* **2018**, *10*, 957. [CrossRef]
7.  Biron, M. *Material Selection for Thermoplastic Parts Practical and Advanced Information for Plastics Engineers*; Elsevier: Oxford, UK, 2016. [CrossRef]
8.  Yuan, W.-J.; Zhao, W.; Wu, G.; Zhao, H.-B. A phosphorus-nitrogen-carbon synergistic nanolayered flame retardant for polystyrene. *Polymers* **2022**, *14*, 2055. [CrossRef] [PubMed]
9.  Hahladakis, J.N.; Velis, C.A.; Weber, R.; Iacovidou, E.; Purnell, P. An overview of chemical additives present in plastics: Migration, release, fate and environmental impact during their use, disposal and recycling. *J. Hazard. Mater.* **2018**, *344*, 179–199. [CrossRef] [PubMed]
10. Wu, Z.; He, C.; Han, W.; Song, J.; Li, H.; Zhang, Y.; Jing, X.; Wu, W. Exposure pathways, levels and toxicity of polybrominated diphenyl ethers in humans: A review. *Environ. Res.* **2020**, *187*, 109531. [CrossRef]
11. Kim, Y.-J.; Osako, M.; Sakai, S.-I. Leaching characteristics of polybrominated diphenyl ethers (PBDEs) from flame-retardant plastics. *Chemosphere* **2006**, *65*, 506–513. [CrossRef] [PubMed]
12. Tian, Z.; Kim, S.-K.; Hyun, J.-H. Environmental distribution of styrene oligomers (SOs) coupled with their source characteristics: Tracing the origin of SOs in the environment. *J. Hazard. Mater.* **2020**, *398*, 122968. [CrossRef] [PubMed]
13. Tawfik, M.S.; Huyghebaert, A. Polystyrene cups and containers: Styrene migration. *Food Addit. Contam.* **1998**, *15*, 592–599. [CrossRef] [PubMed]
14. Auta, H.S.; Emenike, C.U.; Fauziah, S.H. Screening of *Bacillus* strains isolated from mangrove ecosystems in Peninsular Malaysia for microplastic degradation. *Environ. Pollut.* **2017**, *231*, 1552–1559. [CrossRef] [PubMed]
15. Mohanan, N.; Montazer, Z.; Sharma, P.K.; Levin, D.B. Microbial and enzymatic degradation of synthetic plastics. *Front. Microbiol.* **2020**, *11*, 580709. [CrossRef] [PubMed]
16. Kim, H.R.; Lee, H.M.; Yu, H.C.; Jeon, E.; Lee, S.; Li, J.; Kim, D.H. Biodegradation of polystyrene by *Pseudomonas* sp. isolated from the gut of superworms (larvae of *Zophobas atratus*). *Environ. Sci. Technol.* **2020**, *54*, 6987–6996. [CrossRef] [PubMed]
17. Zhang, Y.; Pedersen, J.N.; Eser, B.E.; Guo, Z. Biodegradation of polyethylene and polystyrene: From microbial deterioration to enzyme discovery. *Biotechnol. Adv.* **2022**, *60*, 107991. [CrossRef] [PubMed]
18. Azubuike, C.C.; Chikere, C.B.; Okpokwasili, G.C. Bioremediation techniques-classification based on site of application: Principles, advantages, limitations and prospects. *World J. Microbiol. Biotechnol.* **2016**, *32*, 1461–1466. [CrossRef] [PubMed]
19. Kale, S.K.; Deshmukh, A.G.; Dudhare, M.S.; Patil, V.B. Microbial degradation of plastic: A review. *J. Biochem. Technol.* **2015**, *6*, 952–961.
20. Yuan, J.; Ma, J.; Sun, Y.; Zhou, T.; Zhao, Y.; Yu, F. Microbial degradation and other environmental aspects of microplastics/plastics. *Sci. Total Environ.* **2020**, *715*, 136968. [CrossRef]
21. Tischler, D.; Kaschabek, S.R. Microbial Styrene Degradation: From Basics to Biotechnology. In *Microbial Degradation of Xenobiotics*; Singh, S.N., Ed.; Chapter 3; Springer: Berlin/Heidelberg, Germany, 2012; pp. 67–99. [CrossRef]
22. Nakamiya, K.; Sakasita, G.; Ooi, T.; Kinoshita, S. Enzymic degradation of polystyrene by hydroquinone peroxidase of *Azotobacter beijerinckii* HM121. *J. Ferment. Bioeng.* **1997**, *84*, 480–482. [CrossRef]
23. Hou, L.; Majumder, E.L. Potential for and distribution of enzymatic biodegradation of polystyrene by environmental microorganisms. *Materials* **2021**, *14*, 503. [CrossRef]
24. Amobonye, A.; Bhagwat, P.; Singh, S.; Pillai, S. Plastic biodegradation: Frontline microbes and their enzymes. *Sci. Total Environ.* **2021**, *759*, 143536. [CrossRef]
25. Ali, S.S.; Elsamahy, T.; Al-Tohamy, R.; Zhu, D.; Mahmoud, Y.A.; Koutra, E.; Metwally, M.A.; Kornaros, M.; Sun, J. Plastic wastes biodegradation: Mechanisms, challenges and future prospects. *Sci. Total Environ.* **2021**, *780*, 146590. [CrossRef]
26. Mohan, A.J.; Sekhar, V.C.; Bhaskar, T.; Madhavan Nampoothiri, K. Microbial assisted high impact polystyrene (HIPS) degradation. *Bioresour. Technol.* **2016**, *213*, 204–207. [CrossRef]
27. Kučić Grgić, D.; Miloloža, M.; Lovrinčić, E.; Kovačević, A.; Cvetnić, M.; Ocelić Bulatović, V.; Prevarić, V.; Bule, K.; Ukić, Š.; Markić, M.; et al. Bioremediation of MP-polluted waters using bacteria *Bacillus licheniformis*, *Lysinibacillus massiliensis*, and mixed culture of *Bacillus* sp. and *Delftia acidovorans*. *Chem. Biochem. Eng. Q.* **2021**, *35*, 205–224. [CrossRef]

28. Atiq, N.; Ahmed, S.; Ishtiaq, A.M.; Andleeb, S.; Ahmad, B.; Robson, G. Isolation and identification of polystyrene biodegrading bacteria from soil. *Afr. J. Microbiol. Res.* **2010**, *4*, 1537–1541. Available online: http://www.academicjournals.org/ajmr (accessed on 17 July 2022).

29. Kundu, D.; Hazra, C.; Chatterjee, A.; Chaudhari, A.; Mishra, S. Biopolymer and biosurfactant-graft-calcium sulfate/polystyrene nanocomposites: Thermophysical, mechanical and biodegradation studies. *Polym. Deg. Stab.* **2014**, *107*, 37–52. [CrossRef]

30. Mor, R.; Sivan, A. Biofilm formation and partial biodegradation of polystyrene by the actinomycete *Rhodococcus ruber*. *Biodegradation* **2008**, *19*, 851–858. [CrossRef]

31. Motta, O.; Proto, A.; De Carlo, F.; De Caro, F.; Santoro, E.; Brunetti, L.; Capunzo, M. Utilization of chemically oxidized polystyrene as co-substrate by filamentous fungi. *Int. J. Hyg. Environ. Health* **2009**, *212*, 61–66. [CrossRef] [PubMed]

32. Kumar, A.G.; Hinduja, M.; Sujitha, K.; Rajan, N.N.; Dharani, G. Biodegradation of polystyrene by deep-sea *Bacillus paralicheniformis* G1 and genome analysis. *Sci. Total Environ.* **2021**, *774*, 145002. [CrossRef]

33. Kim, H.-W.; Jo, J.H.; Kim, Y.-B.; Le, T.-K.; Cho, C.-W.; Yun, C.-H.; Chi, W.S.; Yeom, S.-J. Biodegradation of polystyrene by bacteria from the soil in common environments. *J. Hazard. Mater.* **2021**, *416*, 126239. [CrossRef]

34. Shabbir, S.; Faheem, M.; Ali, N.; Kerr, P.G.; Wang, L.-F.; Kuppusamy, S.; Li, Y. Periphytic biofilm: An innovative approach for biodegradation of microplastics. *Sci. Total Environ.* **2020**, *717*, 137064. [CrossRef]

35. Rao, R.S.; Kumar, C.G.; Prakasham, R.S.; Hobbs, P.J. The Taguchi methodology as a statistical tool for biotechnological applications: A critical appraisal. *Biotechnol. J.* **2008**, *3*, 510–523. [CrossRef]

36. Joutey, N.T.; Bahafid, W.; Sayel, H.; Maataoui, H.; Errachidi, F.; Ghachtouli, N.E. Use of experimental factorial design for optimization of hexavalent chromium removal by a bacterial consortium: Soil microcosm bioremediation. *Soil Sediment Contam.* **2014**, *24*, 129–142. [CrossRef]

37. Anthony, J. Full Factorial designs. In *Design of Experiments for Engineers and Scientists*, 2nd ed.; Anthony, J., Ed.; Elsevier: Oxford, MS, USA, 2014; Chapter 6; pp. 63–85. [CrossRef]

38. Ravanipour, M.; Kalantary, R.R.; Mohseni-Bandpi, A.; Esrafili, A.; Farzadkia, M.; Hashemi-Najafabadi, S. Experimental design approach to the optimization of PAHs bioremediation from artificially contaminated soil: Application of variables screening development. *J. Environ. Health* **2015**, *13*, 22. [CrossRef]

39. Briški, F.; Kopčić, N.; Ćosić, I.; Kučić, D.; Vuković, M. Biodegradation of tobacco waste by composting: Genetic identification of nicotine-degrading bacteria and kinetic analysis of transformations in leachate. *Chem. Pap.* **2012**, *66*, 1103–1110. [CrossRef]

40. Benson, H.J. *Microbiological Applications, A Laboratory Manual in General Microbiology*, 8th ed.; The McGraw-Hill Companies: New York, NY, USA, 2001; pp. 154–156, 160, 168, 169.

41. Kyaw, B.M.; Champakalakshmi, R.; Kishore Sakharkar, M.; Lim, C.S.; Sakharkar, K.R. Biodegradation of low density polyethene (LDPE) by *Pseudomonas* species. *Indian J. Microbiol.* **2012**, *52*, 411–419. [CrossRef]

42. Black, J.G. *Microbiology: Principles and Explorations*, 8th ed.; John Wiley & Sons: Hoboken, NJ, USA, 2012; p. 975.

43. *ISO 11348-3*; Water Quality—Determination of the Inhibitory Effect of Water Samples on the Light Emission of *Vibrio fischeri* (*Luminescent bacteria* Test)—Part 3: Method Using Freeze-Dried Bacteria. ISO Publishing: Geneva, Switzerland, 2007.

44. Schmidt-Rohr, K. Oxygen is the high-energy molecule powering complex multicellular life: Fundamental corrections to traditional bioenergetics. *ACS Omega* **2020**, *5*, 2221–2233. [CrossRef]

45. Price, D.; Horrocks, A.R. Combustion processes of textile fibres. In *Handbook of Fire Resistant Textiles*; Woodhead Publishing Limited: Cambridge, UK, 2013; pp. 3–25. [CrossRef]

46. El-Arabi, T.F.; Griffiths, M.W. Bacillus cereus. In *Foodborne Infections and Intoxications*, 4th ed.; Morris, J.G., Morris, E., Eds.; Elsevier: Oxford, MS, USA, 2013; Chapter 29; pp. 401–407. [CrossRef]

47. Yuan, Z.; Nag, R.; Cummins, E. Human health concerns regarding microplastics in the aquatic environment-From marine to food systems. *Sci. Total Environ.* **2022**, *823*, 153730. [CrossRef]

48. Čačko, S.; Pančić, E.; Zokić, I.; Miloloža, M.; Kučić Grgić, D. Aditivi u plastici—Potencijalno štetni učinci na ekosustav. *Chem. Ind.* **2022**, *71*, 49–56. [CrossRef]

49. Gandara e Silva, P.P.; Nobre, C.R.; Resaffe, P.; Seabra Pereira, C.D.; Gusmao, F. Leachate from microplastics impairs larval development in brown mussels. *Water Res.* **2016**, *106*, 364–370. [CrossRef]

50. Suzuki, M.; Suzuki, S.; Matsui, M.; Hiraki, Y.; Kawano, F.; Shibayama, K. Genome sequence of a strain of the human pathogenic bacterium *Pseudomonas alcaligenes* that caused bloodstream infection. *Genome Announc.* **2013**, *1*, e00919-13. [CrossRef]

51. Elkarmi, A.; Abu-Elteen, K.; Khader, M. Modeling the biodegradation efficiency and growth of *Pseudomonas alcaligenes* utilizing 2,4-dichlorophenol as a carbon source pre- and post-exposure to UV radiation. *Jordan J. Biol. Sci.* **2008**, *1*, 7–11.

52. Bisutti, I.; Hilke, I.; Raessler, M. Determination of total organic carbon—An overview of current methods. *TrAC Trends Anal. Chem.* **2004**, *23*, 716–726. [CrossRef]

53. Tian, X.; Zhao, C.; Ji, X.; Feng, T.; Liu, Y.; Bian, D. The correlation analysis of TOC and CODCr in urban sewage treatment. *ICBTE* **2019**, *136*, 06010. [CrossRef]

54. Baker, A.; Cumberland, S.; Hudson, N. Dissolved and total organic and inorganic carbon in some British rivers. *Area* **2008**, *40*, 117–127. [CrossRef]

55. Keller, B.O.; Sui, J.; Young, A.B.; Whittal, R.M. Interferences and contaminants encountered in modern mass spectrometry. *Anal. Chim. Acta* **2008**, *624*, 71–81. [CrossRef]

56. Wang, S.-C.; Gao, Z.-Y.; Liu, F.-F.; Chen, S.-Q.; Liu, G.-Z. Effects of polystyrene and triphenyl phosphate on growth, photosynthesis and oxidative stress of *Chaetoceros meülleri*. *Sci. Total Environ.* **2021**, *797*, 149180. [CrossRef]
57. García Ibarra, V.; Sendón, R.; García-Fonte, X.-X.; Paseiro Losada, P.; Rodríguez Bernaldo de Quirós, A. Migration studies of butylated hydroxytoluene, tributyl acetylcitrate and dibutyl phthalate into food simulants. *J. Sci. Food Agric.* **2019**, *99*, 1586–1595. [CrossRef]
58. De-la-Torre, G.E.; Dioses-Salinas, D.C.; Pizarro-Ortega, C.I.; Saldaña-Serrano, M. Global distribution of two polystyrene-derived contaminants in the marine environment: A review. *Mar. Pollut. Bull.* **2020**, *161*, 111729. [CrossRef]
59. Jung, M.R.; Horgen, F.D.; Orski, S.V.; Rodriguez, V.; Beers, K.L.; Balazs, G.H.; Jones, T.T.; Work, T.M.; Brignac, K.C.; Royer, S.-J.; et al. Validation of ATR FT-IR to identify polymers of plastic marine debris, including those ingested by marine organisms. *Mar. Pollut. Bull.* **2008**, *127*, 704–716. [CrossRef]
60. Subramani, M.; Sepperumal, U. FTIR analysis of bacterial mediated chemical changes in polystyrene foam. *Ann. Biol. Res.* **2016**, *7*, 55–61.
61. Bode, H.B.; Zeeck, A.; Jendrossek, D. Physiological and chemical investigations into microbial degradation of synthetic poly(cis-1,4-isoprene). *Appl. Environ. Microbiol.* **2000**, *66*, 3680–3685. [CrossRef] [PubMed]
62. Mooney, A.; Ward, P.G.; O'Connor, K.E. Microbial degradation of styrene: Biochemistry, molecular genetics, and perspectives for biotechnological applications. *Appl. Microbiol. Biotechnol.* **2006**, *72*, 1–10. [CrossRef] [PubMed]
63. Mao, Y.; Ai, H.; Chen, Y.; Zhang, Z.; Zeng, P.; Kang, L.; Li, W.; Gu, W.; He, Q.; Li, H. Phytoplankton response to polystyrene microplastics: Perspective from an entire growth period. *Chemosphere* **2018**, *208*, 59–68. [CrossRef] [PubMed]
64. Miloloža, M.; Bule, K.; Ukić, Š.; Cvetnić, M.; Bolanča, T.; Kušić, H.; Ocelić Bulatović, V.; Kučić Grgić, D. Ecotoxicological determination of microplastic toxicity on algae *Chlorella* sp.: Response surface modeling approach. *Water Air Soil Pollut.* **2021**, *232*, 327. [CrossRef]

*Article*

# PLA/PHB-Based Materials Fully Biodegradable under Both Industrial and Home-Composting Conditions

Mária Fogašová [1], Silvestr Figalla [2,*], Lucia Danišová [1], Elena Medlenová [1], Slávka Hlaváčiková [1], Zuzana Vanovčanová [1], Leona Omaníková [1], Andrej Baco [1], Vojtech Horváth [1], Mária Mikolajová [1], Jozef Feranc [1], Ján Bočkaj [1], Roderik Plavec [1], Pavol Alexy [1], Martina Repiská [1], Radek Přikryl [2], Soňa Kontárová [2], Anna Báreková [3], Martina Sláviková [3], Marek Koutný [4], Ahmad Fayyazbakhsh [4] and Markéta Kadlečková [4]

[1]  Institute of Natural and Synthetic Polymers, Faculty of Chemical and Food Technology, Slovak University of Technology, Radlinského 9, 812 37 Bratislava, Slovak Republic
[2]  Institute of Materials Science, Faculty of Chemistry, Brno University of Technology, Purkyňova 464/118, 612 00 Brno, Czech Republic
[3]  Department of Landscape Engineering, Hortyculture and Landscape Engineering Faculty, Slovak University of Agriculture, Hospodárska 7, 949 76 Nitra, Slovak Republic
[4]  Department of Environmental Protection Engineering, Faculty of Technology, Tomas Bata University in Zlín, Nad Ovčírnou III 3685, 760 01 Zlín, Czech Republic
*   Correspondence: xcfigallas@vutbr.cz

**Citation:** Fogašová, M.; Figalla, S.; Danišová, L.; Medlenová, E.; Hlaváčiková, S.; Vanovčanová, Z.; Omaníková, L.; Baco, A.; Horváth, V.; Mikolajová, M.; et al. PLA/PHB-Based Materials Fully Biodegradable under Both Industrial and Home-Composting Conditions. *Polymers* 2022, 14, 4113. https://doi.org/10.3390/polym14194113

Academic Editors: Piotr Bulak and Beom Soo Kim

Received: 5 September 2022
Accepted: 23 September 2022
Published: 30 September 2022

**Publisher's Note:** MDPI stays neutral with regard to jurisdictional claims in published maps and institutional affiliations.

**Abstract:** In order to make bioplastics accessible for a wider spectrum of applications, ready-to-use plastic material formulations should be available with tailored properties. Ideally, these kinds of materials should also be "home-compostable" to simplify their organic recycling. Therefore, materials based on PLA (polylactid acid) and PHB (polyhydroxybutyrate) blends are presented which contain suitable additives, and some of them contain also thermoplastic starch as a filler, which decreases the price of the final compound. They are intended for various applications, as documented by products made out of them. The produced materials are fully biodegradable under industrial composting conditions. Surprisingly, some of the materials, even those which contain more PLA than PHB, are also fully biodegradable under home-composting conditions within a period of about six months. Experiments made under laboratory conditions were supported with data obtained from a kitchen waste pilot composter and from municipal composting plant experiments. Material properties, environmental conditions, and microbiology data were recorded during some of these experiments to document the biodegradation process and changes on the surface and inside the materials on a molecular level.

**Keywords:** polylactic acid (PLA); polyhydroxybutyrate (PHB); blend polymeric material; biodegradation; industrial compost; home-compost

## 1. Introduction

Biodegradable polymers are one of the possible alternatives to conventional polymeric materials that can, for specific applications, provide the benefit of biological decomposition without leaving unwanted litter or potentially dangerous microplastics [1,2]. However, biodegradable polymers and materials based on them are significantly different in the conditions needed for their biodegradation and in the time frame of their biodegradation [3–5]. Ideally, for a given application, we need a material that fulfils the necessary application properties and most important mechanical properties and at the same time can biodegrade under the conditions related to the particular application. It is often difficult to find or develop a material that fulfils all of those requirements.

Mainly with the aim to achieve suitable processing and mechanical and barrier properties for a given application, real polymer materials are often blends of several polymers and contain various additives and fillers [6–8]. The whole material has to be biodegradable, which means that correct selection of individual components is very important and can significantly affect its biodegradability. The plasticizer triacetin, used in a study by Sedničková et al. [9], was tested with PLA and PLA/PHB blends, and the materials were exposed to biodegradation in compost at 58 °C. The results of the study prove higher sensitivity of PHB (polyhydroxybutyrate) towards biodegradation in comparison with that of PLA under the same conditions. Additionally, the plasticizer triacetin degraded faster in comparison with PLA. The study also showed that changes in material composition (e.g., amount of plasticizer) might change the biodegradation rate. Another bio-based plasticizer, acetyl-tri-n-butyl citrate (ATBC), in combination with polyethylene glycol (PEG) was also tested as a component of PLA/PHB blends [10]. The material exhibits disintegration under composting conditions in less than one month. The ability of PHB to act as a nucleating agent in PLA/PHB blends slowed down the disintegration, while plasticizer content accelerated it.

It is expected in many cases that waste containing biodegradable plastics is collected along with other primarily plant-based organic waste and it is further treated in municipal or agricultural composting plants. In general, materials suitable for such an end-of-life should comply with EN 13432 standard describing procedures for testing under so-called industrial composting conditions [11]. According to this standard, industrial composting requires an elevated temperature (55–60 °C) in combination with relatively high water activity expressed as water content (approximately 60% $w/w$) and the presence of oxygen. Under such conditions, several criteria must be met: (i) The disintegration of the material from at least 90% must take place within 12 weeks, (ii) 90% mineralization of the composted material must be achieved in less than six months, which is usually measured from evolved $CO_2$, and (iii) the material should not have a negative effect on compost quality (no or minimal heavy metal content, no ecotoxicity) [12].

The description above, however, does not correspond to the conditions on a typical simple composting plant where temperatures over 50 °C can be achieved for about two to three weeks maximally [13]. Then, the process continues in a milder mesophilic temperature range. To address this issue, a group of standards was formulated describing so-called "home-composting" (Vicotte OK compost HOME, AS 5810 Austrian standard) [14]. Here, besides other requirements, the temperature should be kept between 20 and 30 °C during the biodegradation test. Despite the fact that some commercially used materials declared as compostable polymers are often used in applications where composting is meant to be the final stage of life of these materials, they do not meet these requirements and therefore are not compostable. Chemical and biological processes, in general, are, to some extent, accelerated with temperature [15,16]. For example, PLA needs temperatures well over 50 °C to initiate the biodegradation process, which is possibly related to the glass transition temperature of PLA occurring in approximately the same temperature range [17]. This fact was demonstrated, e.g., by Sedničková et al. [9] in a study, where the biodegradation of PLA was measured in compost incubations at 25, 37, and 58 °C for 119 days. The mineralization achieved at 58 °C was 92.3%, but only 19.5% at 37 °C and 14.9% at 25 °C. Other studies showed no or limited PLA biodegradation under mesophilic conditions [18,19].

Moreover, in reality, the almost complete mineralization must be achieved within six months, most of the time at relatively mild temperatures, to make the material compatible with the typical composting plant operation settings [12]. As a consequence, today, typical relatively simple composting plants with no fundamental process control possibilities often tend to reject biodegradable polymer materials, even those labelled as compostable, because, according to their experiences, these items do not decompose fast enough, complicate operations at the plant, and contaminate the resulting compost [20,21].

Another prospective material, PHB, also produced from renewable resources [22,23], can, in contrast to PLA, reach 100% mineralization in five weeks under mild conditions, e.g., in the soil [24]. This polymer itself does not have sufficient processing and mechanical properties for many applications.

The main aim of this study is to demonstrate the biodegradation of the original bio-based biodegradable materials and some model products made of these materials under conditions corresponding to industrial composting (58 °C) and some of them even under conditions corresponding to home-composting (28 °C) in a reasonable time frame. The samples were also evaluated in a real municipal composting plant. Other supporting methods were performed to follow the biodegradation process. This study is meant to show that the presented materials with very good mechanical and processing properties are also biodegradable and fully compatible with common composting technology in a simple municipal composting plant or even in the proper home-compost.

## 2. Materials and Methods

### 2.1. Materials

Samples made of PLA/PHB or PLA/PHB/TPS (thermoplastic starch) blends for this study were developed for various processing technologies and applications. Thermoplastic starch TPS was prepared by blending corn starch + glycerol 70/30 in a twin screw extruder at 160 °C. The used plasticizers are based on esters of citric acid. Blends according to composition in Table 1 were prepared by blending in a co-rotating twin screw extruder with the following parameters: L/D = 44, D = 26 mm, temperature profile from hopper to head: 50-160-170-170-170-170-170-170-160-160 °C, screw speed 150 rpm. The blends were cooled in a water bath and pelletized. All blends were sent to production companies that made the test products. The basic characteristics of all the used blends are listed (Table 1).

**Table 1.** Basic characteristics of the tested materials.

| Blend No. | Blend Technology | PLA Ingeo 4043D %wt. | PHB Enmat Y1000 %wt. | ATBC Citrofol B2 %wt. | TPS %wt. | MFI | P | TS | ε | $T_g$ |
|---|---|---|---|---|---|---|---|---|---|---|
| IM 2 | Injection moulding | 40 | 50 | 10 | 0 | 5.8 | 1.2 | 31 | 8 | N/A |
| IM 1 | Injection moulding | 30 | 40 | 5 | 25 | 35 | 1.3 | 34 | 5 | 53 |
| TF -1 | Thermoforming | 65 | 30 | 5 | 0 | 6.2 | 1.2 | 28 | 36 | N/A |
| FB 2 | Film blowing | 70 | 15 | 15 | 0 | 18 | 1.2 | 18 | 330 | 28 |
| FB 1 | Film blowing | 50 | 10 | 15 | 25 | 33 | 1.3 | 11 | 288 | 24 |

MFI, melt flow index, 180 °C, 2.16 kg, g/10 min; P, density g/cm$^3$; TS, tensile strength at break, MPa; ε, elongation at break, %; $T_g$, glass transition temperature, °C.

The test specimens based on PLA/PHB or PLA/PHB/TPS are listed in Table 2. Cups were produced by the company KS-PT s.r.o. (Slovakia), the thermoforming sheet was produced in Panara a.s. (Slovakia), and blown films were produced in Topstav s.r.o. (Slovakia) in their pilot (Panara a.s.) plant or production plant (KMS-PT s.r.o. and Topstav s.r.o.). The smaller picture of the sample represents the specific used test specimen for biodegradation testing in home-compost, industrial compost, and an electric composter.

**Table 2.** Products that were tested for biodegradation and compostability.

| Sample | Grade No | Thickness | Description |
|---|---|---|---|
| A, bicomponent cup | IM 2 (Inner layer) ———— IM 1 (Outer layer) | 1–6 mm, average 4 mm | Bicomponent cup |

**Table 2.** *Cont.*

| Sample | Grade No | Thickness | Description |
|---|---|---|---|
| B, 500 mL cup | | | |
| | IM 2 | 1 mm | Cup for beverages and beer, mainly for festivals |
| C, 200 mL cup | | | |
| | IM 1 | 1 mm | Cup for drinks suitable mainly for dining restaurants and canteens |
| D, sheet for thermoforming | | | |
| | TF 1 | 0.35 mm | Semiproduct (sheet) for subsequent thermoforming technology, for various packaging applications |
| E, blown film | | | |
| | FB 2 | 0.04 mm | Blown film, e.g., for bags, suitable for various packaging applications |
| F, blown film | | | |
| | FB 1 | 0.04 mm | Blown film e.g., for bags, mainly for waste bags for biowaste collection |

*2.2. Biodegradation Testing by $CO_2$ Production Quantification*

Composting biodegradation tests were performed according to the adapted and miniaturized ISO 14855 method in 500 mL biometric flasks with septum-equipped stoppers. Mature compost from a nearby municipal composting facility (TSZ Ltd., Zlín, Czech Republic) was used in this part of the study. This test was done at 58 °C for industrial composting and at 28 °C to simulate home-composting conditions. Into each flask, 2.5 g of dry-weight compost, 5 g of perlite, and 1 mL of mineral salt medium were weighed, and the water content of the substrate mixture was eventually adjusted to 60% by the addition of sterile drinking water. One hundred milligrams of the samples were cut into 5 × 5 mm fragments that were placed in each sample flask. For each sample, three flasks plus 4 blank flasks were used. The internal production of $CO_2$ in blank incubations was always subtracted to calculate the net sample mineralization. Headspace gas was sampled at appropriate intervals through the septum with a gas-tight needle and conducted through a capillary into a gas analyzer (UAG, Stanford Instruments, Sunnyvale, CA, USA) to determine the amount of $CO_2$. Biodegradation percentage ($D_t$) was calculated as

$$D_t = \frac{(CO_2)_t - (CO_2)_b}{ThCO_2} \times 10 \tag{1}$$

where $(CO_2)_t$ is the released $CO_2$ by each sample, $(CO_2)_b$ is the $CO_2$ produced by the blank flasks, and ($ThCO_2$) is the theoretical $CO_2$ from the sample. A flash elemental analyzer 1112 (Thermo Fisher Scientific, Waltham, MA, USA) was used to measure the carbon content of the samples.

### 2.3. Compostability Testing in an Electric Composter

A small electric composter GG 02 from the JRK company (Slovakia) was used for biodegradation testing while the samples were incubated together with kitchen waste. The effective volume of the composter was 40–50 liters. The temperature was 65 °C during the whole operation time, except for one hour per day when the temperature increased to 75 °C to ensure the hygienization of the content. The internal stirrer was activated for 20 min during each hour, providing altered mixing sequences in forward and reverse directions. The biodegradation process was initiated according to the user manual using the original ACIDULO® bacteria culture. Samples of PLA/PHB or PLA/PHB/TPS blends were inserted into the composter two weeks after the stabilization of the process in the composter. Each day, 0.5–1.0 kg of kitchen food waste was added to the composter. The samples were weighed before being inserting into the composter. Only one sample was measured for every composition because of technical reasons of the experiments. The content of the composter was removed each week and sieved through a sieve with a mesh size of 2 × 2 mm. Pieces larger than 2 mm (which did not pass through the sieve) were collected from the fraction above the sieve. The collected samples were washed in water, subsequently dried in an air oven for 1 h at 90 °C, and weighed with precision of 0.0001 g. Then, the samples were returned to the composter immediately after weighing. Biodegradation was evaluated as the percentage of disintegration. Microbiology inside the composter was monitored with DNA isolation and sequencing following an already established methodology [25].

### 2.4. Disintegration Testing in a Municipal Composting Plant

The disintegration of samples was tested also in the municipal composting facility of the city of Nitra (Nitra District, Southwest Slovakia) under real conditions of industrial composting. The compost pile consisted of approximately 11 m³ of biodegradable municipal waste (a family house garden and public greenery plant-based waste). Disintegration testing was realized in two independent 12-week-long composting cycles. The first cycle was realized from 6 July 2019 to 27 September 2019, and the second cycle—from 16 July 2020 to 12 October 2020 on a dedicated roofed site. Both cycles followed a certified methodology [26]. The samples were weighed and enclosed in a plastic net with a 2 × 2 mm mesh diameter. The cut samples were inserted to approximately 2/3 of the height of the compost pile (Figure 1).

Right below the samples, probes were placed to measure the temperature and humidity inside the compost pile. The data were continuously monitored and recorded. The samples were inspected every 2–3 weeks; sample packages were carefully removed from the pile, visually inspected, and photographed. Afterwards, they were again inserted to 2/3 of the height of the compost pile. The residues of the samples were dried and weighed at the end of the cycle. Biodegradation was evaluated as the percentage of disintegration. The outside air temperature (in both cycles) was about 22 °C on average.

**Figure 1.** The compost pile and the plastic net with the cut samples.

During the first cycle, only one specimen was tested from each sample. The average value of humidity during the entire monitored period was 39.3% (vol). The average value of inner temperature during the first cycle was 62.6 °C. During the second cycle, two specimens were tested for each sample. The average humidity during the entire monitored period was 35% vol. The average inner temperature during the second cycle was 61.1 °C.

*2.5. Material Characterization Methods*

**SEM microscopy.** Surface changes on the tested films were observed using SEM. The samples were coated by a gold/platinum alloy using Balzers SCD 050 sputtering equipment. TESLA BS 300 was used for the observation of samples composted in an electric composter, and JEOL F 7500 SEM (JEOL, Tokio, Japan) was used for the samples from an industrial city composting plant. Phenom Pro Desktop SEM (Thermo Fisher Scientific, Waltham, MA, USA) was used for laboratory experiments under industrial and home-composting conditions.

**Thermophysical properties' measurement.** Differential scanning calorimetry (DSC) was used for the determination of basic thermophysical properties such as glass transition temperature, crystallization temperature, and melting temperature of samples after 0, 6, and 13 days of composting. The conditions for DSC measurements are in Table 3.

**Table 3.** Conditions for DSC measurements.

| Phase | Ramp | Temperature, °C | Time, min |
|---|---|---|---|
| 1. Conditioning | isothermal | 0 | 1 |
| 2. Heating | 10 °C/min | 200 | 20 |
| 3. Conditioning | isothermal | 200 | 1 |

**Gel permeation chromatography (GPC) measurements.** Samples (5 mg) were dissolved in chloroform (1 mL) at 70 °C and filtered through 0.45 μm polytetrafluoroethylene (PTFE) syringe filters. GPC was performed in a 185 Agilent HPLC series 1100 chromatograph (Santa Clara, CA 95051, United States) with a PLgel mixed-c 5 μm, 7.5 × 300 mm column, with chloroform as the mobile phase. Twelve polystyrene standards (0.2–2000 kDa) were used for calibration.

**3. Results and Discussion**

*3.1. Characterization of the Studied Materials*

PLA/PHB- and PLA/PHB/TPS-based materials and final products (films, sheets, and cups) are described in Tables 1 and 2. The materials were developed to be compostable in industrial compost or even under home-composting conditions while still having favorable processing and service properties and containing an important proportion of PLA, which is probably the most available bio-based polymer but still considered to be non-biodegradable under home-composting conditions. All specimens listed in Table 2 were tested in the electric composter, municipal composting plant, and laboratory tests under industrial composting conditions. Samples in the film form (D, E, F) were also tested for biodegradability in a laboratory test under home-composting conditions. The selection of these samples for home-composting was based on their relatively low thickness and on the assumption that home-composting conditions are less aggressive than those in industrial composting, especially for the PLA component of the materials [27,28].

### 3.2. Biodegradability in the Laboratory Test under Industrial and Home-Composting Conditions

All samples were exposed to the laboratory test under industrial composting conditions (58 °C). It was expected that based on the composition of the samples; they all should be completely mineralized under these conditions. For the thick-wall samples without TPS (A and B), more time to reach 100% mineralization was assumed.

Three samples for each composition were measured, including the reference sample (cellulose). The average standard deviation for all tested compositions was ±8.6 for the industrial composting conditions. Mineralization of 100% was obtained for all tested samples after about 90 days of incubation (Figure 2). No sample except E exhibited a lag phase. Sample E contained the highest amount of PLA (70%), so this typical feature of PLA compost biodegradation was demonstrated in this sample [28,29]. All other samples, which degraded without a lag phase, contained at least 30% of easily biodegradable PHB and/or TPS, which were able to smooth out the lag phase in the biodegradation curve under industrial compost conditions. In the case of E, the PHB phase was probably closed inside the dominant PLA phase. After the lag phase, mineralization went on exponentially, and the sample reached complete mineralization as the first one. The mineralization was also fast for other film samples D and F, with F being faster at the beginning, which probably reflected its higher content of easily biodegradable TPS and plasticizers (40%) in comparison to that in D (5%). Surprisingly, the mineralization was also fast for a relatively thick (1 mm) sample C with high PHB and TPS contents (65%). The slowest degradation was observed in the case of both thick samples A (4 mm) and B (1 mm). TPS-containing cups exhibited slightly faster biodegradation in very good correlation with the cup's construction (thickness). Interestingly, the thickness of the sample, at least to some extent, did not play a very significant role. The thin film F composed of a PLA/PHB/TPS blend exhibited a course of biodegradation similar to that of the thick sample C. Both were made of a comparable formulation, but if TPS was not present in the composition, thickness played a more important role (samples B, D, and E were TPS-free formulations). Possibly, the TPS phase could initiate early disintegration of the sample and thus circumvent the importance of thickness.

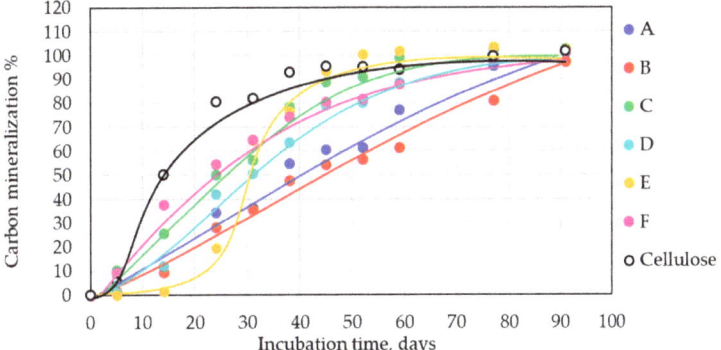

**Figure 2.** Mineralization of samples under industrial composting conditions, 58 °C.

The selected samples were removed from the compost after incubation, and they were observed in SEM to evaluate microbial colonization and deterioration of the material's surface (Figure 3). After only 10 days in the compost, all materials were densely colonized, and the density of the biofilm was clearly increased with the time of exposure. It was not possible to identify the microorganisms present, but the morphological appearance suggested filamentous thermophilic actinobacteria with distinguishable round endospores. The erosion of the surface was also clearly apparent.

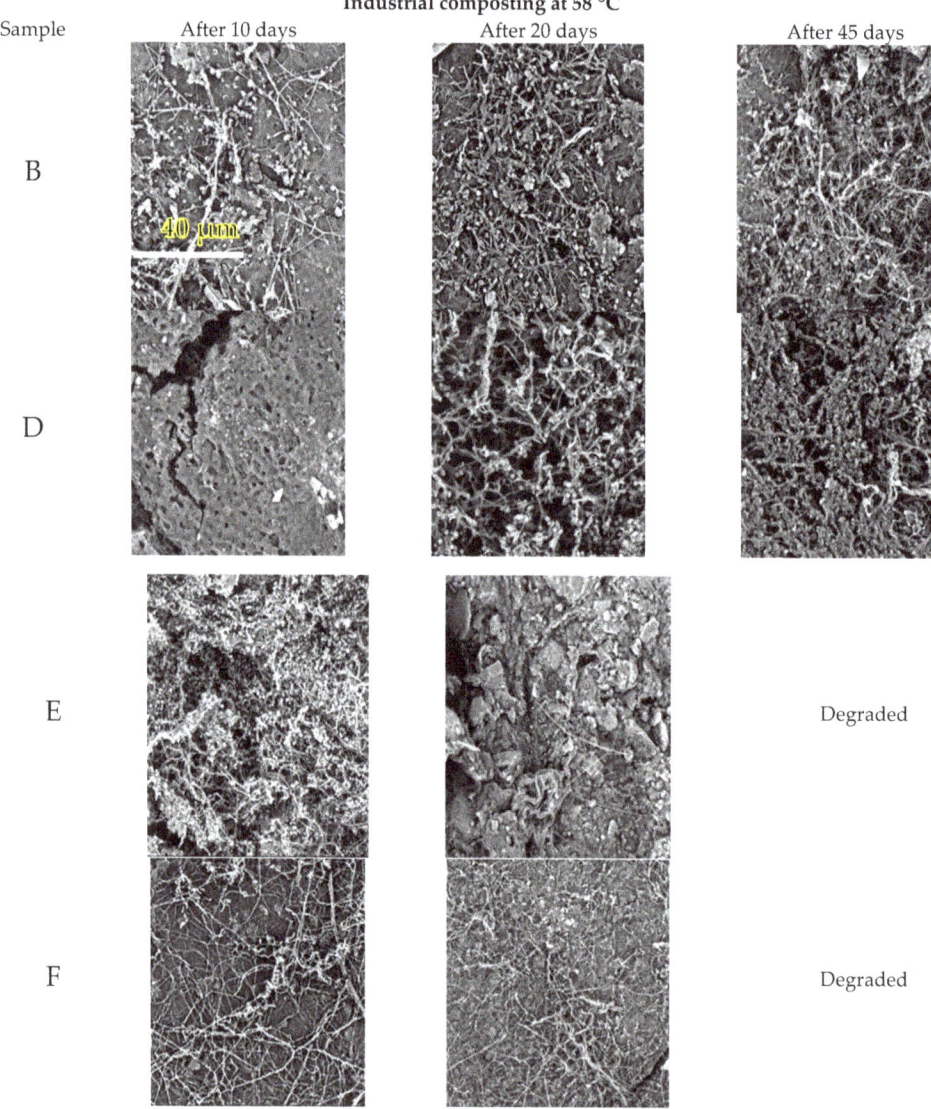

**Figure 3.** Scanning electron microscopy survey of the sample surface during the biodegradation test at 58 °C (industrial composting condition). Magnification of 5000×. Specifications according to Table 2 (B—500 mL cup, D—sheet for thermoforming, E—blown film and F—blown film with TPS).

Film samples D, E, and F were selected for biodegradation under home-composting conditions (28 °C, Figure 4). From our previous experiences [28,29] and the majority of the scientific literature on the topic [30], it should be expected that the PLA fraction of the materials should not be mineralized under such conditions.

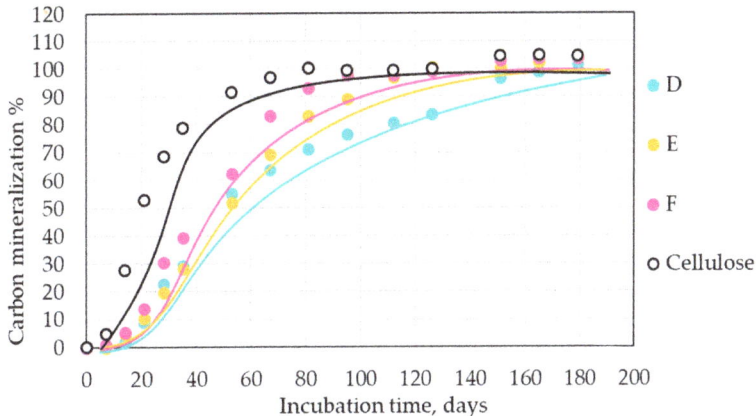

**Figure 4.** Mineralization of samples under home-composting conditions, 28 °C.

Three samples for each composition were measured, including the reference sample (cellulose). All samples exhibited total mineralization in a period of about 180 days. The average standard deviation for all tested compositions was ±3.9 for the home-composting conditions. An about 15-day-long lag phase occurred for all tested samples. Additionally, the curves for the tested samples were not so far from that of the cellulose reference and had the expected order. The thin film of the PLA/PHB/TPS blend (F) exhibited the fastest biodegradability, followed with a minimal gap by the thin film sample without TPS (E). In this case, apparently, thickness and plasticizer played a significant role; the thick film made of PLA/PHB without TPS and at lower plasticizer content (D) degraded significantly slower than did the thin film with higher plasticizer content (E).

When observed in SEM (Figure 5), the samples from the home-compost experiments, in general, showed a much lower degree of surface colonization if compared with that in the industrial compost experiment, which can be explained by the fact that at this temperature, a completely different microbial community was present. Again, filamentous (actinomyces most probably) but also rod-shaped bacteria were discernible (e.g., Figure 5F, 30 days). (1) (2) (3) Cavities and cracks were gradually formed. At this temperature, fungi were very active, with their extracellular enzymes probably degrading the material even if they were not seen attached to the surface.

A very important result from the environmental as well as practical points of view is the fact that PLA/PHB-blend samples degraded fully at home-compost conditions. It means that 100% mineralization was reached despite the general opinion that PLA is not able to biodegrade at temperatures below its $T_g$, (about 55 °C) under home-compost conditions and therefore only the mineralization of the non-PLA part of the blends could be expected. The blending of PLA and PHB in the hot melt state probably causes re-esterification reactions, leading to the formation of PLA/PHB co-polyesters. The easily biodegradable PHB segments in such a co-polyester can promote a release of low MW fragments that are prone to further biodegradation [27]. It is very difficult to investigate the occurrence and extent of re-esterification by standard analytical methods, but the results from the home-composting biodegradation can be considered as proof. This is indeed a significant discovery not only for the eventual composting "at home" but also for industrial composting plants. In these usually municipal facilities, the composting process is quite simple and not always fully controlled, so the thermophilic phase of the composting process could be too brief or not sufficiently hot to initiate total PLA mineralization. Problems have been reported with various items from PLA-based material labelled as "compostable", and often, such items are no longer accepted in these plants. Additionally, this will prevent microplastic formation, one of the most-closely watched environmental risks studied recently.

| Sample | After 30 days | After 55 days | After 80 days |
|---|---|---|---|

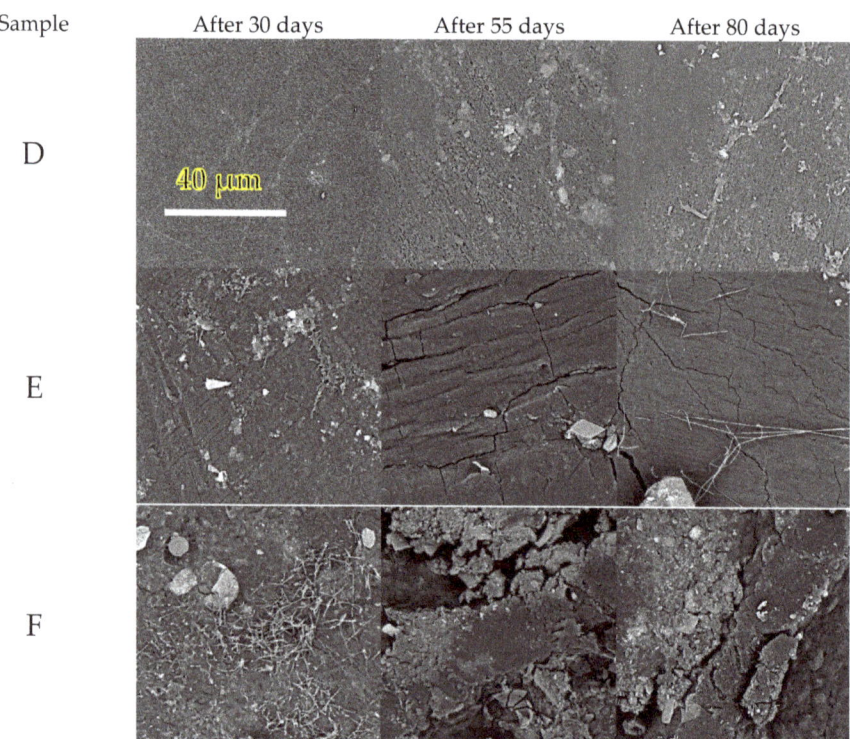

**Figure 5.** Scanning electron microscopy survey of the sample surface during the biodegradation test at 28 °C (home-composting conditions). Magnification of 5000×. Specifications according to Table 2 (D—sheet for thermoforming, E—blown film and F—blown film with TPS).

*3.3. Composting in the Electric Composter*

All samples in Table 2 were also tested in a small electric composter. Test specimens of 10 × 7 cm with the original thickness were inserted into the composter. Biodegradation was evaluated as weight losses in percentages; it can also be stated that the degree of disintegration was evaluated. The results presented in Figure 6 show differences between the disintegration of individual samples. The disintegration of all film samples was very fast, and all films were disintegrated into pieces smaller than 2 mm after 20 days. The cup containing TPS (C) exhibited fast disintegration into particles smaller than 2 mm after 40 days. The other two cups (B, PLA/PHB-based; A, TPS-containing cup) were disintegrated after a very similar time period, while the combined cup A was slightly faster. The comparison of biodegradation curves based on $CO_2$ measurements (in industrial compost conditions and home-compost conditions (Figure 6) and disintegration curves (Figure 6) can provide an important insight into the problem of the eventual formation of microplastics.

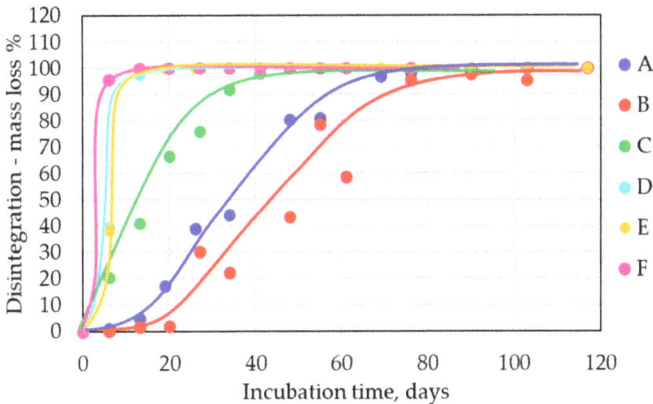

**Figure 6.** The disintegration of tested samples in a small electric composter, expressed as weight loss.

In Figure 7, sample C is shown as a typical example of all three cups. It means that the composting process in this case ensured direct microbial conversion of the materials to $CO_2$ with a relatively short disintegration step. On the contrary, the films rapidly disintegrated into smaller particles, and the mineralization followed immediately. This observation in the case of films was given by the low mechanical strength of the samples and the mechanical strain during the mixing inside the composter. A thick cup was more resistant to mechanical breakdown than thin films were before the samples became too brittle due to the biodegradation process. A similar effect was also observed when home-composting and electric composter biodegradation were compared in the case of sample E (Figure 7), where, logically, a more significant delay of mineralization after disintegration was detected. These results show that the studied materials really also underwent mineralization in home-composting conditions, and eventual fragments/microplastics after their disintegration were readily mineralized.

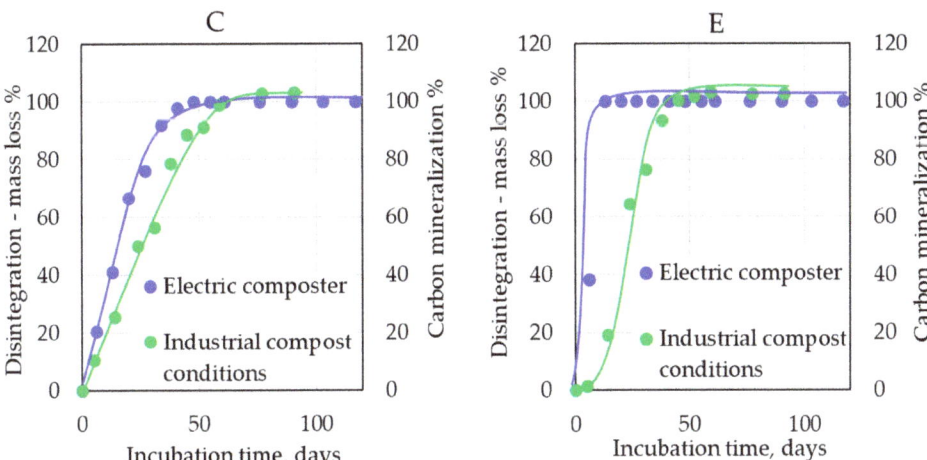

**Figure 7.** Comparison of biodegradation curves measured as $CO_2$ in industrial compost conditions (58 °C) for samples C and E and in home-compost conditions (28 °C) for sample E with disintegration curves obtained using an electric composter for samples C and E.

All tested samples exhibited well-visible changes in the surface morphology after only 6 days of incubation, as can be seen in the SEM figures. Thick samples B without TPS and

C with TPS that were not disintegrated instantly are shown as examples (Figure 8). The microscopic observation is in perfect correlation with the results of biodegradation and disintegration testing discussed previously. Cavities of various dimensions and depths were created in relation to the sample´s composition, which resulted in the enlargement of the surface area and acceleration of the biodegradation process.

500× original          500× after 6 days

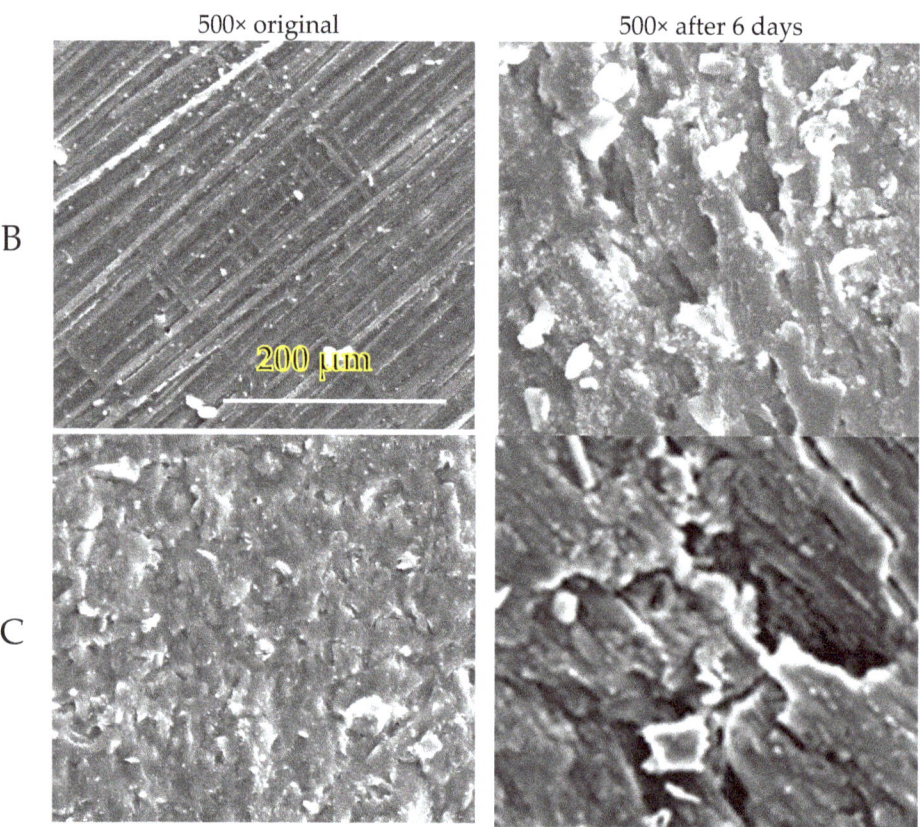

**Figure 8.** SEM images of the tested samples before composting and after 6 days of incubation in an electric composter. Specifications according to Table 2. (B—500 mL cup and C—200 mL cup with TPS).

The composition of the bacterial community was described by next-generation 16S r DNA metagenome sequencing in different stages of the process. The design of the composter and its operation exhibited a severe limitation for the bacterial community. Very high operation temperature (65 °C) and a daily hygienization period (75 °C) put even thermophilic bacteria on the limit of their survival. These parameters should be adjusted to provide a better environment for the compost microflora; however, it is not the topic of this study. Evidently (Figure 9), the community was dominated by thermophilic spore-forming taxa (Bacilli, Actinobacteria). The presence of other taxa like Bacteroidia, Negativicutes, and even Clostridia witness probably the presence of anaerobic pockets inside the compost. The introduced inoculum strongly influenced the initial stage, then, the gradual increase of the other taxa with time could be seen.

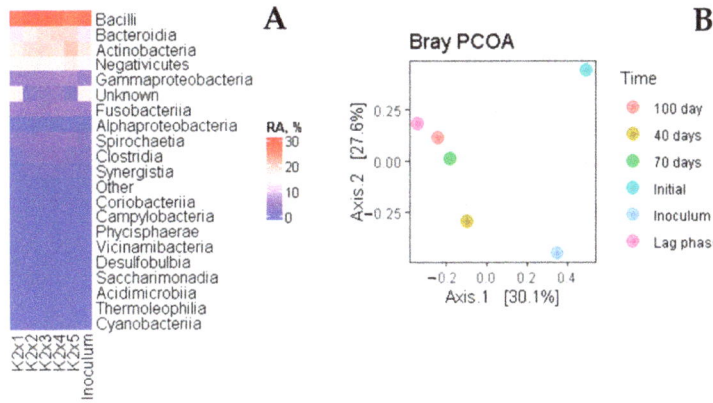

**Figure 9.** The bacterial community inside an electric composter. (**A**): heatmap at class taxonomic level, RA, relative abundance. (**B**): PCoA scatter plot.

### 3.4. Composting at the Municipal Composting Plant

The study of compostability under real conditions and verification of the results from the laboratory and the electric composter were realized in two composting cycles at Nitra municipal composting plant (Slovakia). Samples B, C, and F were tested during the first cycle. All investigated (Table 2) samples were tested in the second cycle. The first and second cycles were realized separately, the first one during the summer of 2019 and the second one during the summer of 2020. The main difference between the first and second cycles was in the humidity curve of the compost substrate (Figure 10). During the first cycle, the composting process started with a high level of moisture which then decreased over time. The moisture profile in the second cycle was exactly the opposite. In both cases, pure pulp paper (pure cellulose) was used as a reference. Biodegradation was evaluated from the weight losses after 12 weeks of incubation.

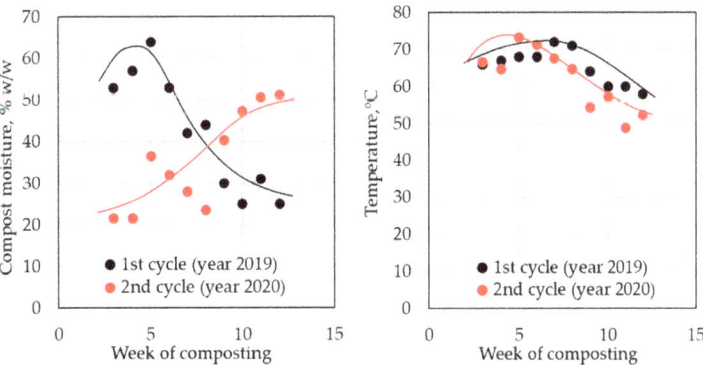

**Figure 10.** Moisture and temperature profiles during composting in a municipal composting plant.

Significant differences in the first cycle (year 2019) were observed in the degradation of the pulp paper reference in both cycles (Figures 11 and 12). While temperature profiles were very similar in both cycles, the trends of the compost substrate humidity differed significantly. The first stages of composting in the second cycle proceeded apparently at insufficient humidity, which could explain why in the second cycle, the compost was not sufficiently active for the biodegradation of the reference material. Suitable moisture content is an essential parameter for the composting process, especially during the initial hot phases of the composting. The pulp paper reference was not found in the first cycle of composting after 12 weeks, while in the second cycle, with low humidity in the first stages of composting, lower than 50% weight loss was observed (Figure 12), and the paper still preserved its original shape (Figure 11).

| Original pulp paper before composting | Pulp paper after 12 weeks in the first composting cycle (year 2019) | Pulp paper after 12 weeks in the second composting cycle (year 2020) |
|---|---|---|

Pulp paper was not found

**Figure 11.** Pulp paper reference material before industrial composting and after 12 weeks of composting.

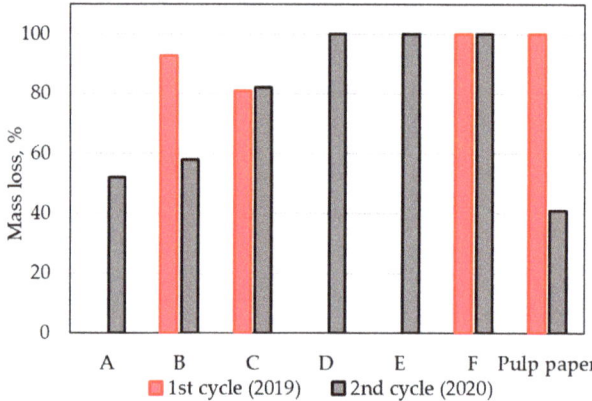

**Figure 12.** Weight losses of the tested samples after 12 weeks of incubation in the first and second composting cycles at a municipal composting plant.

The investigated samples provided very interesting results regarding the above-discussed differences between both composting cycles. Despite the humidity profile, film samples degraded completely in both composting cycles (sample F in the first composting cycle, as well as samples D, E, F in the second composting cycle). A thick cup sample containing TPS (sample C) degraded equally in both cycles, and the humidity profile had a small effect on their biodegradation. The thick sample without TPS (cup sample B) and the combined cup (sample A) were probably more sensitive to the moisture content but still biodegraded substantially. In general, the samples in the study biodegraded comparably or even better than the cellulose reference did in the described real-condition experiment.

The surface morphology of the samples was also studied in the second cycle of the industrial composting experiment. Film samples could not be analyzed because they were decomposed completely; all other samples exhibited significant changes in surface morphology (Figure 13). Similarly, as with the samples from the electric composter, strong surface erosion was observed.

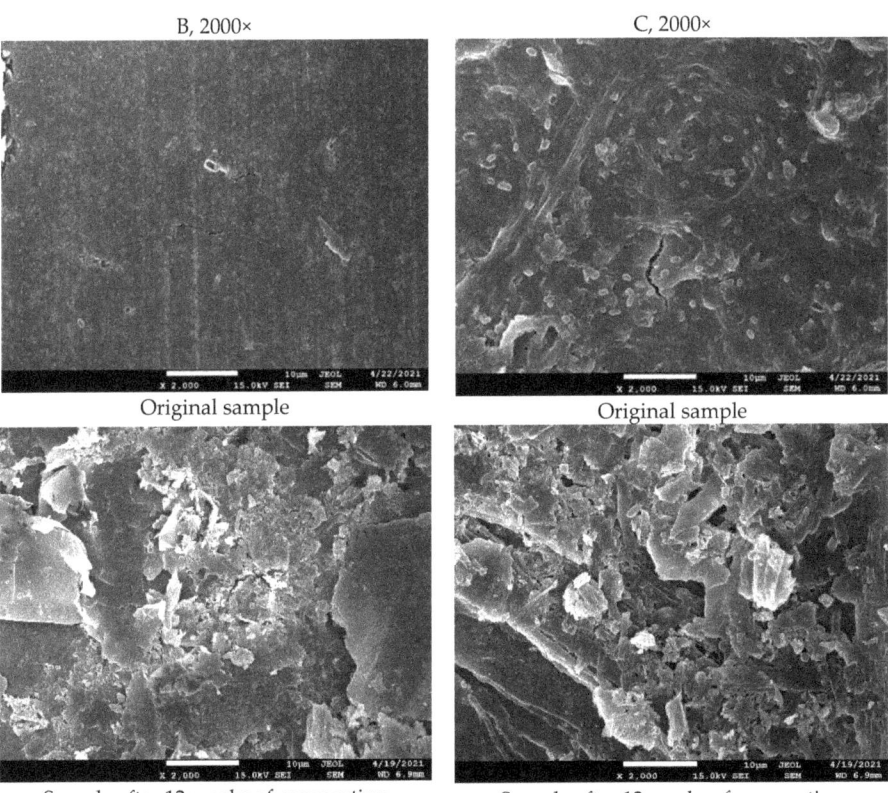

Figure 13. SEM images after 12 weeks of incubation at a municipal composting plant (second cycle).

*3.5. Changes in the Material Properties during the Composting Experiments*

In the case of the electric composter, after each period, molecular weight distributions were measured by GPC. It must be noted that GPC measurements were realized with solid, nondegraded residuals of composted materials. GPC curves of unprocessed PLA and PHB used in the blends were also analyzed, and the records are shown (Figure 13, insert). Partial degradation due to blending and then processing of the blend during sample preparation was noticeable for each sample. The degradation caused by processing only in the melt state was not very extensive; only in the case of sample C was its extent more significant. This can relate to the presence of glycerol in TPS and possible alcoholysis of PHB by glycerol [31,32]. The intense shift of the main peaks to lower molecular weights after composting was visible mainly in the case of thick cup samples. In addition, the appearance of low-molecular-weight fractions was clearly seen in GPC records. The last indicated fraction produced by the biodegradation process exhibited an average molecular weight (MW) of about 1000 g/mol. Lower MW fractions (below 500 g/mol) were probably easily mineralized. The fast gradual decrease of MW was observed for all samples (Figure 14).

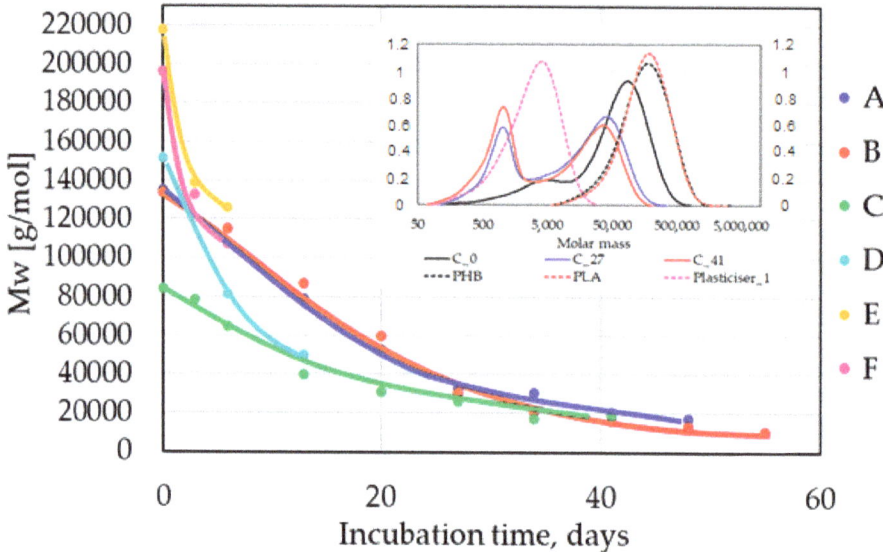

**Figure 14.** MW evolution during incubation in an electric composter. The GPC records of sample C after 0, 27, and 41 days of composting are shown as an example (insert).

Regardless of the original MW, all samples that did not biodegrade exhibited similar MW values after 30 days of composting. It means that the degradation process ran not only on the surface of the samples but also in the bulk of the materials. Only short time periods were evaluated in the case of film samples because already after 14 days of composting, no film sample could be retrieved.

The process at the municipal composting plant also caused a dramatic decline in MW below 1000 g/mol, as shown for selected samples (Figure 15). Such low-molecular-weight substances are considered to be easily and rapidly biodegradable during the following composting period.

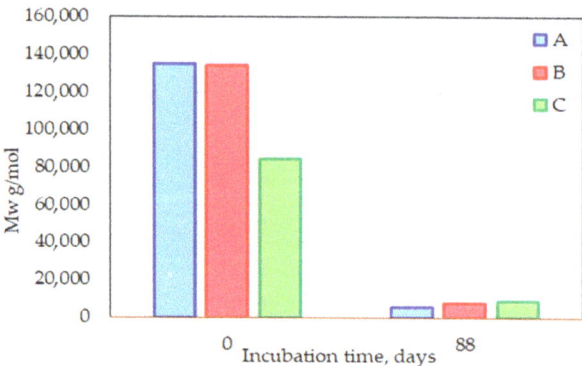

**Figure 15.** Comparison of MW for cup samples before composting and after 12 weeks in the second composting cycle (2020).

The mentioned changes in polymer structure were also confirmed by DSC measurements realized on solid residuals after each evaluated period of composting process in the electric composter (Figure 16).

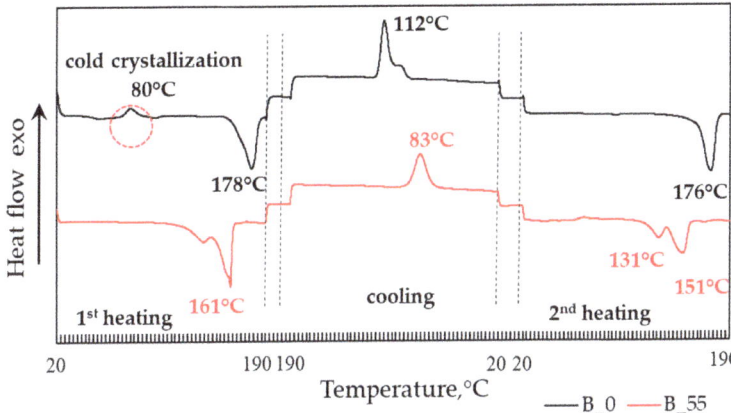

**Figure 16.** DSC curves of sample B before composting (B_0) and after 55 days of composting (B_55) in an electric composter.

Significant changes were visible in the first heating as well as in the cooling and the second heating cycle. The temperatures of melting were significantly shifted to lower values after 55 days of composting; after the same period, the crystallization temperature was significantly decreased. The composting caused the disappearance of the cold crystallization peak visible in the original sample, and the melting peak was split in two. Quantitative analysis of DSC measurements (Figure 17) also showed that the changes in thermograms in the first run were not connected only to the consumption of the amorphous phase, but significant changes in the molecular structure occurred. This was confirmed by the decrease of crystallization enthalpy as well as of the melting enthalpy in the second heating cycle when the record of the sample was no longer affected by its previous thermal and other history.

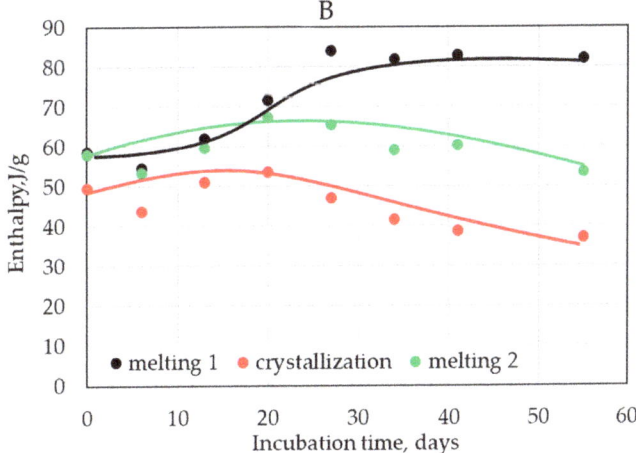

**Figure 17.** Evolution of enthalpies of melting and crystallization during the incubation in an electric composter for sample B.

## 4. Conclusions

This detailed study presents results from composting experiments in several experimental settings, all aimed at the group of PLA/PHB blend materials with various compositions and, depending on the particular material, also containing citric-acid-ester-based plasticizers and thermoplastic starch (TPS). All tested samples were in their real final product forms and shapes.

It was found that the studied PLA/PHB-based blends were fully biodegradable under industrial composting conditions as well as in an electric composter, which was designed for the composting of kitchen waste. A very important result from this study is the observation that some of the studied materials, despite their high PLA contents, could fully biodegrade under home conditions. The result is explained by the assumption that PLA could react with PHB during the blending in the melt form, and this reactive extrusion process could induce re-esterification of both polymer components. This extrusion process was intentionally designed with such a purpose. The easily biodegradable PHB segments can promote the polymer chains' scission and ultimately complete biodegradation of the materials under home-composting conditions.

Evaluation of GPC, DSC, and SEM data showed that during the composting of studied PLA/PHB blends, the changes in the molecular structure and morphology proceeded not only on the surface but also in the bulk of the materials. SEM images showed that already in the first stages of biodegradation, not only macroscopic disintegration but also biological decomposition of the materials took place on the surface, causing an enlargement of the specific surface area and thus the acceleration of the biodegradation process.

A very important result was the verification that disintegration and mineralization as the two main processes during biodegradation of materials could run in the case of the studied PLA/PHB blends not only in sequence but also in parallel. This was observed mainly in the case of thicker specimens like cups, but also in the case of thin products like films where the disintegration was extremely rapid but the mineralization phase was delayed only shortly. Based on these results, it can be concluded that these PLA/PHB materials do not leave microplastics in the environment after industrial as well as after home-composting processes. The composting experiment in a municipal composting plant confirmed and verified the laboratory results. During two independent testing cycles, it was found that the studied PLA/PHB materials degraded comparably or even faster than the pure cellulose reference did.

**Author Contributions:** Conceptualization, R.P. (Roderik Plavecand); methodology, M.F. and M.K. (Marek Koutný); validation, R.P. (Radek Přikryl); formal analysis, E.M., S.H., J.B. and M.R.; investigation, M.F., S.F., L.O., V.H., A.F. and M.S.; data curation, A.B. (Andrej Bacoand) and S.K.; writing—original draft preparation, L.D. and S.F.; visualization, M.M. and A.B. (Anna Báreková); supervision, P.A. and M.K. (Markéta Kadlečková); project ad-ministration, E.M., Z.V. and J.F. All authors have read and agreed to the published version of the manuscript.

**Funding:** This research was funded by Internal Grant Agency with grant number IGA/FT/2022/006, Slovak Research and Development Agency with grant number APVV-20-0256 and APVV-20-0193, Operational Program Integrated Infra-structure within the project "Demand-driven research for the sustainable and innovative food", Drive4SIFood 313011V336, co-financed by the European Regional Development Fund and project FCH-S-21-7553 of the Specific University Research Grant.

**Institutional Review Board Statement:** Not applicable.

**Data Availability Statement:** The study did not report any data.

**Acknowledgments:** This work was supported by an internal grant IGA/FT/2022/006, APVV-20-0256, APVV-20-0193, Operational Program Integrated Infrastructure within the project "Demand-driven research for the sustainable and innovative food", Drive4SIFood 313011V336, co-financed by the European Regional Development Fund and project FCH-S-21-7553 of the Specific University Research Grant.

**Conflicts of Interest:** The authors declare no conflict of interest.

## References

1. Duraj-Thatte, A.M.; Manjula-Basavanna, A.; Courchesne, N.-M.D.; Cannici, G.I.; Sánchez-Ferrer, A.; Frank, B.P.; Hag, L.V.; Cotts, S.K.; Fairbrother, D.H.; Mezzenga, R.; et al. Water-processable, biodegradable and coatable aquaplastic from engineered biofilms. *Nat. Chem. Biol.* **2021**, *17*, 732–738. [CrossRef]
2. Xia, Q.; Chen, C.; Yao, Y.; Li, J.; He, S.; Zhou, Y.; Li, T.; Pan, X.; Yao, Y.; Hu, L. A strong, biodegradable and recyclable lignocellulosic bioplastic. *Nat. Sustain.* **2021**, *4*, 627–635. [CrossRef]
3. Ghosh, K.; Jones, B.H. Roadmap to Biodegradable Plastics—Current State and Research Needs. *ACS Sustain. Chem. Eng.* **2021**, *9*, 6170–6187. [CrossRef]
4. Liao, J.; Chen, Q. Biodegradable plastics in the air and soil environment: Low degradation rate and high microplastics formation. *J. Hazard. Mater.* **2021**, *418*, 126329. [CrossRef]
5. Šašinková, D.; Serbruyns, L.; Julinová, M.; FayyazBakhsh, A.; De Wilde, B.; Koutný, M. Evaluation of the biodegradation of polymeric materials in the freshwater environment—An attempt to prolong and accelerate the biodegradation experiment. *Polym. Degrad. Stab.* **2022**, *203*, 110085. [CrossRef]
6. Arockiam, A.J.; Subramanian, K.; Padmanabhan, R.; Selvaraj, R.; Bagal, D.K.; Rajesh, S. A review on PLA with different fillers used as a filament in 3D printing. *Mater. Today Proc.* **2021**, *50*, 2057–2064. [CrossRef]
7. Ko, E.; Kim, T.; Ahn, J.; Park, S.; Pak, S.; Kim, M.; Kim, H. Synergic Effect of HNT/oMMT Bi-filler System for the Mechanical Enhancement of PLA/PBAT Film. *Fibers Polym.* **2021**, *22*, 2163–2169. [CrossRef]
8. Shanmugam, V.; Rajendran, D.J.J.; Babu, K.; Rajendran, S.; Veerasimman, A.; Marimuthu, U.; Singh, S.; Das, O.; Neisiany, R.E.; Hedenqvist, M.S.; et al. The mechanical testing and performance analysis of polymer-fibre composites prepared through the additive manufacturing. *Polym. Test.* **2020**, *93*, 106925. [CrossRef]
9. Sedničková, M.; Pekařová, S.; Kucharczyk, P.; Bočkaj, J.; Janigová, I.; Kleinová, A.; Jochec-Mošková, D.; Omaníková, L.; Perďochová, D.; Koutný, M.; et al. Changes of physical properties of PLA-based blends during early stage of biodegradation in compost. *Int. J. Biol. Macromol.* **2018**, *113*, 434–442. [CrossRef]
10. Arrieta, M.P.; López, J.; Rayón, E.; Jiménez, A. Disintegrability under composting conditions of plasticized PLA&PHB blends. *Polym. Degrad. Stab.* **2014**, *108*, 307–318. [CrossRef]
11. Rahman, H.; Bhoi, P.R. An overview of non-biodegradable bioplastics. *J. Clean. Prod.* **2021**, *294*, 126218. [CrossRef]
12. Lee, P.-K.; Choi, B.-Y.; Kang, M.-J. Assessment of mobility and bio-availability of heavy metals in dry depositions of Asian dust and implications for environmental risk. *Chemosphere* **2015**, *119*, 1411–1421. [CrossRef]
13. Yu, Z.; Tang, J.; Liao, H.; Liu, X.; Zhou, P.; Chen, Z.; Rensing, C.; Zhou, S. The distinctive microbial community improves composting efficiency in a full-scale hyperthermophilic composting plant. *Bioresour. Technol.* **2018**, *265*, 146–154. [CrossRef]
14. Altieri, R.; Seggiani, M.; Esposito, A.; Cinelli, P.; Stanzione, V. Thermoplastic Blends Based on Poly(Butylene Succinate-co-Adipate) and Different Collagen Hydrolysates from Tanning Industry—II: Aerobic Biodegradation in Composting Medium. *J. Polym. Environ.* **2021**, *29*, 3375–3388. [CrossRef]
15. Mengqi, Z.; Shi, A.; Ajmal, M.; Ye, L.; Awais, M. Comprehensive review on agricultural waste utilization and high-temperature fermentation and composting. *Biomass Convers. Biorefinery* **2021**, *24*, 1–24. [CrossRef]
16. Schulte, P.M. The effects of temperature on aerobic metabolism: Towards a mechanistic understanding of the responses of ectotherms to a changing environment. *J. Exp. Biol.* **2015**, *218*, 1856–1866. [CrossRef]
17. Iovino, R.; Zullo, R.; Rao, M.A.; Cassar, L.; Gianfreda, L. Biodegradation of poly(lactic acid)/starch/coir biocomposites under controlled composting conditions. *Polym. Degrad. Stab.* **2008**, *93*, 147–157. [CrossRef]
18. Apinya, T.; Sombatsompop, N.; Prapagdee, B. Selection of a Pseudonocardia sp. RM423 that accelerates the biodegradation of poly(lactic) acid in submerged cultures and in soil microcosms. *Int. Biodeterior. Biodegrad.* **2015**, *99*, 23–30. [CrossRef]
19. Narancic, T.; Verstichel, S.; Chaganti, S.R.; Morales-Gamez, L.; Kenny, S.T.; De Wilde, B.; Padamati, R.B.; O'Connor, K.E. Biodegradable Plastic Blends Create New Possibilities for End-of-Life Management of Plastics but They Are Not a Panacea for Plastic Pollution. *Environ. Sci. Technol.* **2018**, *52*, 10441–10452. [CrossRef]
20. Accinelli, C.; Abbas, H.K.; Bruno, V.; Nissen, L.; Vicari, A.; Bellaloui, N.; Little, N.S.; Shier, W.T. Persistence in soil of microplastic films from ultra-thin compostable plastic bags and implications on soil Aspergillus flavus population. *Waste Manag.* **2020**, *113*, 312–318. [CrossRef]
21. Bandini, F.; Frache, A.; Ferrarini, A.; Taskin, E.; Cocconcelli, P.S.; Puglisi, E. Fate of Biodegradable Polymers Under Industrial Conditions for Anaerobic Digestion and Aerobic Composting of Food Waste. *J. Polym. Environ.* **2020**, *28*, 2539–2550. [CrossRef]
22. Koskimäki, J.J.; Kajula, M.; Hokkanen, J.; Ihantola, E.-L.; Kim, J.H.; Hautajärvi, H.; Hankala, E.; Suokas, M.; Pohjanen, J.; Podolich, O.; et al. Methyl-esterified 3-hydroxybutyrate oligomers protect bacteria from hydroxyl radicals. *Nat. Chem. Biol.* **2016**, *12*, 332–338. [CrossRef] [PubMed]
23. Opgenorth, P.H.; Korman, T.P.; Bowie, J.U. A synthetic biochemistry module for production of bio-based chemicals from glucose. *Nat. Chem. Biol.* **2016**, *12*, 393–395. [CrossRef] [PubMed]
24. Siracusa, V.; Rocculi, P.; Romani, S.; Rosa, M.D. Biodegradable polymers for food packaging: A review. *Trends Food Sci. Technol.* **2008**, *19*, 634–643. [CrossRef]
25. Šerá, J.; Serbruyns, L.; De Wilde, B.; Koutný, M. Accelerated biodegradation testing of slowly degradable polyesters in soil. *Polym. Degrad. Stab.* **2020**, *171*, 109031. [CrossRef]

26. Vaverková, M.; Adamcová, D.; Kotovicová, J.; Toman, F. Evaluation of biodegradability of plastics bags in composting conditions. *Ecol. Chem. Eng. S* **2014**, *21*, 45–57. [CrossRef]
27. Husárová, L.; Pekařová, S.; Stloukal, P.; Kucharzcyk, P.; Verney, V.; Commereuc, S.; Ramone, A.; Koutny, M. Identification of important abiotic and biotic factors in the biodegradation of poly(l-lactic acid). *Int. J. Biol. Macromol.* **2014**, *71*, 155–162. [CrossRef]
28. Stloukal, P.; Kalendova, A.; Mattausch, H.; Laske, S.; Holzer, C.; Koutny, M. The influence of a hydrolysis-inhibiting additive on the degradation and biodegradation of PLA and its nanocomposites. *Polym. Test.* **2015**, *41*, 124–132. [CrossRef]
29. Stloukal, P.; Pekařová, S.; Kalendova, A.; Mattausch, H.; Laske, S.; Holzer, C.; Chitu, L.; Bodner, S.; Maier, G.; Slouf, M.; et al. Kinetics and mechanism of the biodegradation of PLA/clay nanocomposites during thermophilic phase of composting process. *Waste Manag.* **2015**, *42*, 31–40. [CrossRef]
30. Jašo, V.; Glenn, G.; Klamczynski, A.; Petrović, Z.S. Biodegradability study of polylactic acid/ thermoplastic polyurethane blends. *Polym. Test.* **2015**, *47*, 1–3. [CrossRef]
31. Janigová, I.; Lacík, I.; Chodák, I. Thermal degradation of plasticized poly(3-hydroxybutyrate) investigated by DSC. *Polym. Degrad. Stab.* **2002**, *77*, 35–41. [CrossRef]
32. Špitalský, Z.; Lacík, I.; Lathová, E.; Janigová, I.; Chodák, I. Controlled degradation of polyhydroxybutyrate via alcoholysis with ethylene glycol or glycerol. *Polym. Degrad. Stab.* **2006**, *91*, 856–861. [CrossRef]

*Article*

# Assessment of a Coated Mitomycin-Releasing Biodegradable Ureteral Stent as an Adjuvant Therapy in Upper Urothelial Carcinoma: A Comparative In Vitro Study

Federico Soria [1,*], Salvador David Aznar-Cervantes [2], Julia E. de la Cruz [1], Alberto Budia [3], Javier Aranda [1], Juan Pablo Caballero [4], Álvaro Serrano [5] and Francisco Miguel Sánchez Margallo [1]

[1] Jesus Uson Minimally Invasive Surgery Centre Foundation, Endoscopy-Endourology Department, 10071 Cáceres, Spain; jecruz@ccmijesususon.com (J.E.d.l.C.); javierarandap@gmail.com (J.A.); msanchez@ccmijesususon.com (F.M.S.M.)
[2] Biotechnology Department, IMIDA, 30150 Murcia, Spain; sdac1@um.es
[3] Urology Department, Polytechnic and University La Fe Hospital, 46026 Valencia, Spain; alberto.budia@hotmail.com
[4] Urology Department, University General Hospital, 03010 Alicante, Spain; juanpablocaballero@gmail.com
[5] Urology Department, University Clinico San Carlos, 28040 Madrid, Spain; alvaro.serrano.p@gmail.com
* Correspondence: fsoria@ccmijesususon.com

**Citation:** Soria, F.; Aznar-Cervantes, S.D.; de la Cruz, J.E.; Budia, A.; Aranda, J.; Caballero, J.P.; Serrano, Á.; Sánchez Margallo, F.M. Assessment of a Coated Mitomycin-Releasing Biodegradable Ureteral Stent as an Adjuvant Therapy in Upper Urothelial Carcinoma: A Comparative In Vitro Study. *Polymers* 2022, 14, 3059. https://doi.org/10.3390/polym14153059

Academic Editor: Piotr Bulak

Received: 7 June 2022
Accepted: 26 July 2022
Published: 28 July 2022

**Abstract:** A major limitation of the treatment of low-grade upper tract urothelial carcinoma is the difficulty of intracavitary instillation of adjuvant therapy. Therefore, the aim of this in vitro study was to develop and to assess a new design of biodegradable ureteral stent coated with a silk fibroin matrix for the controlled release of mitomycin C as a chemotherapeutic drug. For this purpose, we assessed the coating of a biodegradable ureteral stent, BraidStent®, with silk fibroin and subsequently loaded the polymeric matrix with two formulations of mitomycin to evaluate its degradation rate, the concentration of mitomycin released, and changes in the pH and the weight of the stent. Our results confirm that the silk fibroin matrix is able to coat the biodegradable stent and release mitomycin for between 6 and 12 h in the urinary environment. There was a significant delay in the degradation rate of silk fibroin and mitomycin-coated stents compared to bare biodegradable stents, from 6–7 weeks to 13–14 weeks. The present study has shown the feasibility of using mitomycin C-loaded silk fibroin for the coating of biodegradable urinary stents. The addition of mitomycin C to the coating of silk fibroin biodegradable stents could be an attractive approach for intracavitary instillation in the upper urinary tract.

**Keywords:** ureteral stent; biodegradable stent; drug eluting stent; chemotherapy; UTUC

## 1. Introduction

Urothelial carcinomas are the sixth most common tumors in developed countries. They may be located in the lower urinary tract, bladder or urethra, as well as in the upper urinary tract, ureter, and pyelocaliceal system. Bladder tumors are responsible for 90–95% of urothelial carcinomas and are the most common neoplasm of the urinary tract. Upper tract urothelial carcinomas (UTUC) are uncommon and represent only 5–10% of total cases, with a yearly incidence of almost 2 cases per 100,000 population [1].

Patients with UTUC may have one of two types of urothelial carcinoma: low-grade or high-grade. It is very useful to stratify the risk of UTUC into low- and high-grade in order to identify those patients who will benefit from nephron-sparing surgery and those who should instead undergo radical treatment, such as nephroureterectomy [2].

In order to categorize a tumor as low risk, each of the following characteristics must be met: unifocal disease, tumor size <2 cm, negative urine cytology for high-grade tumor, positive biopsy for low-grade tumor, and non-invasive appearance on a computed tomography (CT) scan [3]. These patients may benefit from minimally invasive treatment using a transurethral endoscopic approach with a semi-rigid or flexible endoscope. Additionally, an adjuvant instillation of mitomycin C (MMC) has provided promising results in previous studies and may reduce the risk of urothelial recurrence and progression in patients affected by low grade UTUC [3,4]. Unfortunately, there is currently no suitable procedure for adjuvant chemotherapeutic intracavitary instillation after holmium laser fulguration of UTUC tumor lesions [5]. This is due to the difficulty of intracavitary chemotherapy instillation in the upper urinary tract, given the washout effect of urine production at the renal level and the low storage capacity of the upper urinary tract [6]. As a result, many patients do not benefit from an adjuvant chemotherapy instillation procedure, which leads to worse results in this group of patients.

A recent meta-analysis highlighted that novel drug delivery technologies promise to change this paradigm by favoring drug exposure in the upper tract, leading to higher treatment efficacy [7]. One of the new technologies for chemotherapy instillation in the upper urinary tract is a gel loaded with 4 mg/mL MMC. When it is instilled, it is liquid, but at body temperature it gels and is removed within 4–6 h, through the urinary tract. It is used for primary chemo-ablation. A single-arm, open-label, phase III clinical trial has shown very encouraging results, which have been evaluated up to 12 months. This system was approved by the Food and Drug Administration (FDA, Silver Spring, MD, USA) 2 years ago; it shows a successful response in 59% of patients but 27% experience severe side effects, and the administration schedule is very complicated for patients [8].

A further new development in the delivery of cytostatics to the upper urinary tract comes from vascular stents, called 'drug eluting stents' [9]. Our research group has developed a biodegradable ureteral stent known as the BraidStent® [10]. This platform has been coated with a multi-layered silk fibroin (SF) protein coating for the controlled release of MMC. SF has previously demonstrated outstanding properties as a drug eluting coating, such as for stent applications, based on its useful mechanical properties and biological outcomes [11–13]. The scientific literature has high expectations for fibroin, reporting that it represents a new biomaterial with improved mechanical properties as a scaffold. It is expected to overcome the limitations of current biomaterials for stent coating, both as a polymer carrier and because of its demonstrated biocompatibility and easy processability. In fact, many studies have shown that SF is more biocompatible than other currently-used polymeric degradable biomaterials, such as PLGA (poly-lactic-co-glycolic acid), PLA (polylactic acid), and PGA (polyglycolic acid) [11,14,15].

The aim of this comparative in vitro experimental study was to develop and to assess a new design of biodegradable ureteral stent coated with a SF matrix for the controlled release of MMC for intracavitary instillation in the adjuvant treatment of UTUC.

## 2. Materials and Methods

The experimental study was organized into three protocols.

**Experimental protocol I**. In the first protocol, the aim was to compare two combinations of biodegradable polymers and copolymers for the manufacture of the biodegradable ureteral stent before coating it with SF.

**Materials for stent preparation.** Three polymers and copolymers were selected for this purpose: Glycomer™ 631 (Biosyn suture by Covidien, Minneapolis, MN, USA), PGA (Safil® Quick suture by B. Braun, Secaucus, NJ, USA), and poly-4-hydroxybutyrate (Monomax® suture by B. Braun Surgical, Barcelona, Spain). All three biomaterials are derived from biocompatible and biodegradable surgical sutures. The search for the right combination of polymers for the biodegradable ureteral stent aimed to produce a stent that degrades within 7–8 weeks in a progressive manner to avoid obstructive degradation in future in vivo studies.

In order to compare the degradation rate, 3-cm long fragments of biodegradable ureteral stents were developed using the following combinations: Group BraidStent-1, a long-term braided stent with Glycomer™ 631 and poly-4-hydroxybutyrate; and group BraidStent-2, a short-term braided stent with Glycomer™ 631 and PGA. The ratio in the composition of the polymers of each stent was always kept constant in their manufacture: Glycomer™ 631 (54%); PGA and poly-4-Hydroxybutyrate (46%).

Five samples of each of the types of stent fragments were developed, giving a total of 10 samples. These were placed in watertight tubes with artificial urine (human synthetic urine, BioIVT, Royston, UK), pH: 6.7 specific gravity 1.008 (FDA registered) for screening studies for high performance liquid chromatography with diode-array detection (HPLC-DAD). To investigate the degradation rate, the stent fragments were dipped in 5 mL of artificial urine (AU) and incubated in an orbital shaker-incubator under mimicked biological conditions (36.5 °C with 5% $CO_2$, at 90 rpm) until complete stent biodegradation, with daily AU changes. Four follow-ups were performed in each group: T0—start of the study; T1—day of onset of macroscopic degradation; T2—day on which we macroscopically detected that the stent had degraded by 50%; and T3—complete stent degradation. The changes in pH, manifestation of nitrites, weight of the wet stent, and days at which the described changes appeared were assessed. The average thickness of each stent was also determined at baseline (T0) (Figure 1).

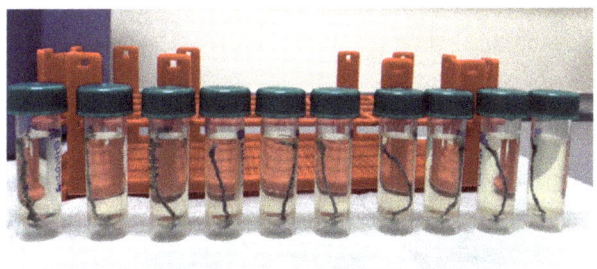

**Figure 1.** Ureteral stents fragments dipping in artificial urine (Day 6).

**Experimental protocol II.** Once the study corresponding to protocol I had been completed and the most suitable polymer/copolymer combination for the development of the BraidStent® was known, we proceeded to the next step, which consisted of the SF coating of the BraidStent® (the BraidStent-SF group).

**Materials for stent preparation and SF coating.** Cocoons of *Bombyx mori* were obtained from worms reared in the sericulture facilities of the IMIDA, Biotechnology Department (Murcia, Spain). Cocoons were chopped up and boiled in 0.02 M $Na_2CO_3$ for 30 min in order to eliminate the sericin. Then, the raw SF was rinsed with distilled water and dried at room temperature for 3 days. Subsequently, SF was dissolved in 9.3 M LiBr (Acros Organics) for 3 h at 60 °C, yielding a 20% $w/v$ dissolution that was dialyzed against distilled water for 3 days (Snakeskin Dialysis Tubing 3.5 KDa MWCO, Thermo Scientific, Waltham, MA, USA) with 8 total water changes (at 4 °C) [16]. The resultant 7–8% $w/v$ SF solution was recovered and used for the preparation of the coated BraidStent-SF stents as explained below, adjusting the concentration to 7% $w/v$ before use. The protocol was adapted from the methodology proposed by Rockwood et al. for the manufacture of fibroin tubes by means of a dipping technique using alternate baths of aqueous fibroin and methanol [17].

The BraidStent-SF group was composed of the same number of stent fragments (five) and followed the same experimental protocol, with the same follow-ups and assessment of the same variables as in protocol I (Figure 2).

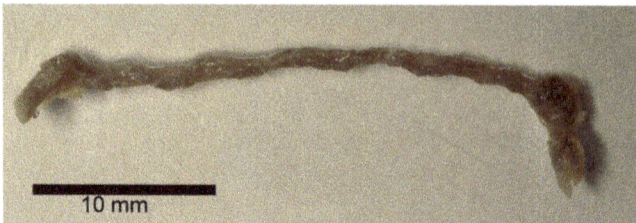

**Figure 2.** BraidStent-SF sample. The SF coating of the stent can be appreciated.

**Characterization**. In order to assess the appropriate SF coating on the stent surface, a comparative study was carried out with three samples per group, where BraidStent-2 was compared to BraidStent-SF with sulforhodamine B (SRB) (a fluorescent aqueous marker) coating added to visualize the homogeneity of the SF coating by fluorescence microscopy.

**Experimental protocol III**. Following the evaluation of the BraidStent-SF, it was loaded with two different MMC formulations (BraidStent-SF-MMC1 and BraidStent-SF-MMC2) to assess the release concentration and MMC release time. The BraidStent-SF-MMC1 and 2 groups were composed of the same number of stent fragments, 5 per group, and followed the same experimental protocol and follow-ups as in protocols I and II.

**Materials for stent preparation and SF and MMC coating.** For the BraidStent-SF-MMC1 group, the coating of the stents used powdered 10 mg MMC to produce the intravenous injectable solution commonly used in the medical field (INIBSA, Barcelona, Spain); the vials contained sodium chloride as an excipient (9.5 mg of sodium per 1 mg of MMC). For this, a 7% $w/v$ fibrin solution containing 5 mg/mL of MMC was prepared by dissolving the drug for 30 min under orbital agitation at 120 rpm. At the same time, the same protocol was performed using absolute methanol, dissolving MMC at the same concentration and for the same period of time. These two starting solutions were used to alternately bathe the stent fragments, by dipping for 5 s in the first solution (of aqueous fibroin) and then in the second (containing methanol), for the same time, and letting them dry slightly for 1 min before repeating the procedure. These coating cycles were carried out in this BraidStent-SF-MMC1 group 10 times. The containers used for this purpose were 5 mL glass tubes (Figure 3).

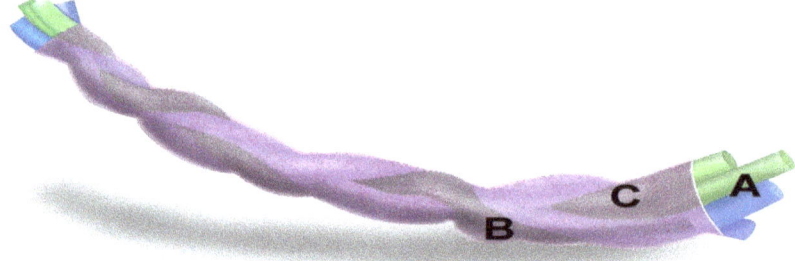

**Figure 3.** Illustration of BraidStent-SF-MMC. **A**—BraidStent; **B**—SF coating; **C**—MMC embedded in SF matrix.

For the BraidStent-SF-MMC2 group, the concentration employed for both the fibroin and methanol solutions was 10 mg/mL, and 10 dip coating cycles were performed. In this group, 70 mg of pure MMC (Mitomycin CRS, Sigma-Aldrich, Steinheim am Albuch, Germany) was used, without any excipient (Figure 3).

The stents were completely stable in room air. However, the MMC coating needed special care, i.e., it must be packaged in a lightproof for preservation and transportation to ensure its physical and chemical stability. MMC shows some light sensitivity, so reasonable steps to minimize light exposure should be taken.

**MMC assessment.** MMC concentrations were assessed every six hours until there was no analytical evidence of its detection. Urine from each follow-up was replaced and analyzed. MMC was released from the SF matrix due to solubility events of the SF matrix in the urinary environment. The permeability and release kinetics of the MMC depends on the SF coating and relates to the percentage of the beta-sheet structure. Increasing the crystallinity (methanol immersion) of the SF beta sheet decreases the release rate and increases the release duration.

The HPLC-DAD method was an isocratic method with a mobile phase of acetonitrile: ultrapure water 80:20 ($v/v$) at a flow rate of 1 mL/min, and separation was performed on a LUNA C18 250 mm, 4.6 mm, 5 μm column at 30 °C temperature. A total of 10 μL of sample was injected. The MMC was detected with a diode array detector (DAD) at 365 nm (1260 Infinity II Prime LC System, Agilent Technologies, Santa Clara, CA, USA).

**Statistical analysis.** Statistical analysis was performed with the SPSS 25.0 program for Windows (IBM, Armonk, NY, USA). The variables studied were pH, weight of the stents in g, degradation rate expressed in days, stent width in mm, and artificial urine concentration of MMC in mg/L. Variables are shown as their mean ± standard deviation. The normality of the data was analyzed using the Shapiro–Wilk test. For data that followed a normal distribution, the intra-group and intra-phase distributions were analyzed using a one-factor ANOVA. Post-hoc analysis was performed using Tukey's HSD (honestly significant difference) test.

For variables in which the data were not normally distributed, the comparison between groups in each phase was carried out using the Kruskal–Wallis test. In the event of statistical significance, the corresponding post-hoc pairwise comparison was carried out. The trends of the variables throughout the follow-ups concerning the effect of the factors time and group, were analyzed by means of a General Linear Model (GLM) Repeated Measures. Again, the post-hoc analysis was performed using Tukey's HSD. In certain cases, to evaluate the trend of each group along time, a Friedman test or a Wilcoxon test was performed, depending on the number of follow-ups included in the analysis. In addition, the concentration of Mitomycin C released by the two groups, BraidStent-SF-MMC1 and BraidStent-SF-MMC2, was analyzed either via a t-test for independent samples or a Mann–Whitney test, depending on the normality of data. Confidence intervals were set at 95% and significance was determined by $p$-values less than 0.05.

## 3. Results

**Protocol I.** We found significant differences between the two groups for the variable 'stent weight' at T0 and T1. The combined weight of polymers and copolymers in BraidStent-1 was significantly higher than in BraidStent-2. BraidStent-1 showed a weight loss of 11.11% and BraidStent-2 showed a weight loss of 14.81% between the baseline study and the macroscopic onset of degradation (Figure 4). The duration of degradation variable shows a statistically significant difference between both groups at T1 and T3, but not at T2 (Figure 5). BraidStent-2 started degrading later, but between T2 and T3, there was an acceleration of the hydrolysis process, resulting in a shorter time to complete stent degradation. BraidStent-2 fit the criterion of degradation in the first 7–8 weeks (Figure 5). The difference in the pH of the medium in protocol I showed statistical significance at T1 and T2. The stent degradation metabolites of BraidStent-1 and BraidStent-2 caused overt acidification of the medium, which was more marked with BraidStent-1. At T2 and T3, the pH returned nearly to basal levels. Using Friedman's statistical test, we assessed the trend over the different phases and whether the changes in pH within each group were uniform (Figure 6). In both groups, the pH trend was not uniform and showed significance throughout the different phases (BraidStent-1 group $p = 0.006$ and BraidStent-2 group $p = 0.009$). None of the samples were positive for nitrites on urinalysis.

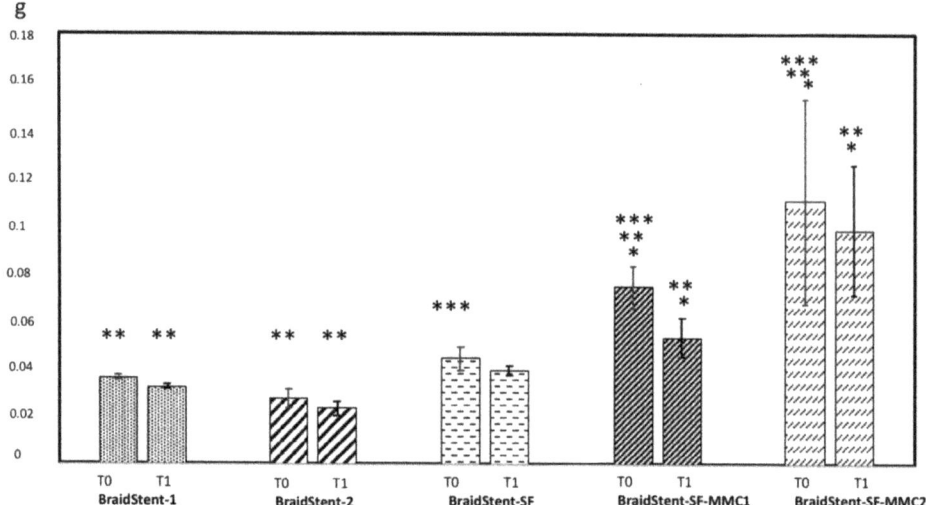

**Figure 4.** Assessment of weight loss (grams) between the start of the study (T0) and the start of stent degradation (T1). Values indicate mean ± SD. Significance of the values between each tested groups at T0 and T1 was determined by the Kruskal–Wallis H test and post hoc pairwise comparison. Wilcoxon test was used to analyze weight loss from T0 to T1 within each group. Significant differences are expressed as: intra-groups (* $p < 0.05$); inter-groups (** $p < 0.05$) (BraidStent-1 vs. BraidStent-2; BraidStent-SF-MMC1 vs. BraidStent-SF-MMC2); inter-groups (*** $p < 0.05$) (BraidStent-SF vs. BraidStent-SF-MMC1 and BraidStent-SF-MMC2).

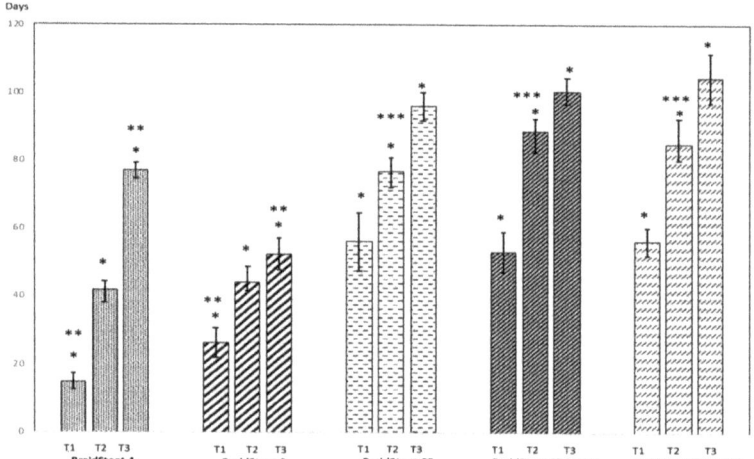

**Figure 5.** Degradation rate assessment (days) between T1 (onset of degradation); T2 (50% of stent degradation); T3 (complete stent degradation). Values indicate mean ± SD. Significance of the values among tested groups within each follow-up was determined by the Kruskal–Wallis H test and post hoc pairwise comparison. The effect of group and time along follow-ups from T1 to T3 was assessed via a GLM Repeated Measures and pairwise comparison via Tukey's test. Statistical significance is depicted as follows: intra-groups (* $p < 0.05$); inter-groups (** $p < 0.05$) (BraidStent-1 vs. BraidStent-2; BraidStent-SF-MMC1 vs. BraidStent-SF-MMC2); inter-groups (*** $p < 0.05$) (BraidStent-SF vs. BraidStent-SF-MMC1 and BraidStent-SF-MMC2).

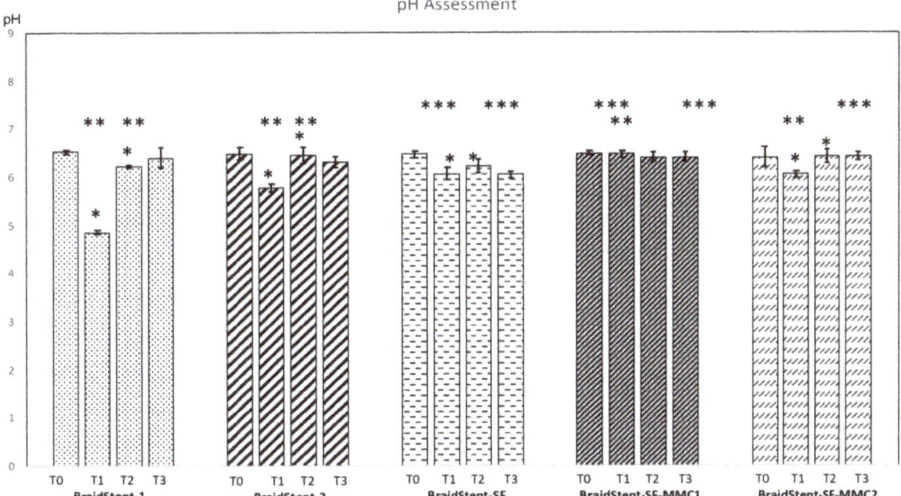

**Figure 6.** Assessment of pH throughout the study until complete stent degradation. T0 (start of study); T1 (onset of degradation); T2 (50% of degradation); T3 (complete stent degradation). Values indicate mean ± SD. Significance of the values between each tested groups within each phase of the study was determined by the Kruskal–Wallis H test and post hoc pairwise comparison; while a General Linear Model (GLM) Repeated Measures was used to evaluate the effect of both group and time (from T0 to T3) in the trend of pH, with the corresponding post-hoc analysis by Tukey test. Statistical significance is indicated in the figure as follows: intra-groups (* $p < 0.05$); inter-groups (** $p < 0.05$) (BraidStent-1 vs. BraidStent-2; BraidStent-SF-MMC1 vs. BraidStent-SF-MMC2); inter-groups (*** $p < 0.05$) (BraidStent-SF vs. BraidStent-SF-MMC1 and BraidStent-SF-MMC2). Friedman test shows that only the BraidStent-SF-MMC1 group is homogeneous along follow-ups.

**Protocol II.** For this protocol, the BraidStent-2 group was selected as it met the inclusion criteria of complete degradation before 8 weeks. The BraidStent-SF group used the same combination of polymers and copolymers as in the BraidStent 2 group, dip-coated with SF. The characterization results of the comparative study between BraidStent-2 and BraidStent-SF SRB coated show, after evaluation with fluorescence microscopy, that the SF coating of the stent is uniform (Figure 7).

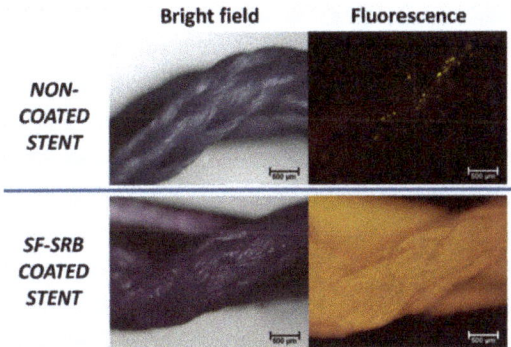

**Figure 7.** Assessment of SF coating with SRB fluorescent dye. Comparative study between BraidStent-2 (non-coated stent) and BraidStent-SF with SRB coating. Fluorescence microscopy showed adequate homogeneity of SF coating.

The BraidStent-SF group did not show statistically significant changes between T0 and T1, with a decrease in the weight of 11.36% between the two phases (Figures 4 and 8). With regard to degradation time, the addition of the SF coating led to significant differences compared to the BraidStent groups, significantly increasing the degradation time compared to the stent without the SF coating at T1-T2-T3 (Kruskal–Wallis H test, $p = 0.002$). The Friedman test determined that the trend in pH between the study phases showed statistical significance ($p = 0.016$). As in groups BraidStent-1 and -2, there was acidification of the urinary medium at the beginning of degradation, which subsequently recovered to basal levels at T3 (Figure 6). None of the samples were positive for nitrites on urinalysis, ruling out bacterial contamination.

**10 mm**

**Figure 8.** BraidStent-SF sample in T2. It can be appreciated that the SF coating has practically disappeared.

**Protocol III.** Following the addition of the two MMC formulations to the stent, the weight showed statistically significant changes between the BraidStent-SF-MMC1 and the BraidStent-SF-MMC2 groups, with a greater weight in the latter group at T0 and T1. Both groups showed statistically significant differences compared to the SF-coated stent without MMC; thus, the addition of MMC significantly increased the weight of the stent (Figure 4). Moreover, stent thickness showed statistically significant differences between the BraidStent-SF group and the two groups coated with MMC (Figure 9). There was no statistically significant difference between the BraidStent-SF-MMC1 and the BraidStent-SF-MMC2 groups with regard to the manifestation of the different degradation phases, however, when both groups were compared with BraidStent-SF, statistical significance was found at T2, with this latter group showing a much faster degradation at this follow-up (Figure 5). The BraidStent-SF-MMC1 group, in contrast to BraidStent-SF-MMC2, shows a uniform trend in its variations throughout the urinary pH study (Friedman's test, $p = 0.357$) (Figure 6). Both stent groups underwent the same number of coating cycles, 10, although the initial concentration of MMC in BraidStent-SF-MMC1 was 10 mg compared to 70 mg in BraidStent-SF-MMC2. In addition, the joint dilution of SF and MMC was 5 mg/mL and 10 mg/mL, respectively. The kinetics of MMC in both groups was fully released within the first 12 h, with no analytical evidence of MMC afterward. We found statistically significant differences in the concentration of MMC released at 12 h between the groups, with a higher urine concentration of MMC in the BraidStent-SF-MMC2 group (Table 1). Statistically significant differences were also found within each group between 6 and 12 h. None of the samples were positive for nitrites on urinalysis.

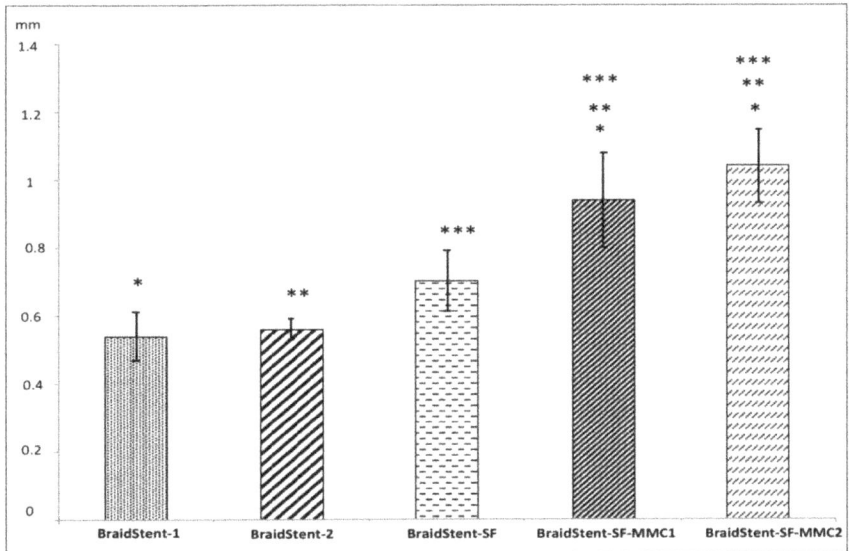

**Figure 9.** Stent thickness in the experimental groups at baseline (T0) (mm). Values indicate mean ± SD. Significance of the values between each tested group was determined by a one-way ANOVA and a post hoc Tukey test: (* $p < 0.05$) (BraidStent-1 versus the rest of experimental groups); (** $p < 0.05$) (BraidStent-2 versus the rest of experimental groups); (*** $p < 0.05$) (BraidStent-SF versus the rest of experimental groups).

**Table 1.** Mitomycin C concentration release rate (mg/L). MMC release is assessed every 6 h in artificial urine in BraidStent-SF-MMC1 and BraidStent-SF-MMC2 stents, until it is not detected by HPLC-DAD. The variable mg/L MMC at 6 h follows a normal distribution but not at 12 h. A Student's *t*-test for independent samples was performed in order to determine the differences at 6 h between groups. As for the study at 12 h, a Mann–Whitney test was performed. The trend in MMC concentration over the 6 and 12 h, between groups and within each group, was analyzed using a GLM Repeated measures. Superscripts refer to statistical significance between the values, namely inter- or intra-group. a-b-c ($p < 0.01$).

| mg/L | 6 h | 12 h | 18 h | 24 h |
|---|---|---|---|---|
| BraidStent-SF-MMC1 | 10.98 ± 3.73 [a] | 2.33 ± 3.05 [ac] | 0 | 0 |
| BraidStent-SF-MMC2 | 7.67 ± 1.39 [b] | 56.08 ± 4.76 [bc] | 0 | 0 |

## 4. Discussion

In response to the clear need to design new intracavitary chemotherapy instillation systems for adjuvant treatment of low-grade UTUC, gels and ureteral stents that can deliver chemotherapeutics have been developed [7–9,18,19]. MMC-eluting gels are already FDA-approved; these represent an important advance, despite having demonstrated a high complication rate in the first clinical study [8,18]. There is also an in vitro study on a chemotherapy-eluting ureteral stent reported in the scientific literature [9,19]. These two developments for intracavitary instillation of chemotherapy present important limitations at the clinical level, since, in the case of the gel, it is necessary to introduce a ureteral catheter for instillation, and in the case of the ureteral stent, it is necessary to remove it cystoscopically. The design we assessed in this experimental study uses a biodegradable ureteral stent able to release MMC as a chemotherapeutic agent. This avoids the necessary removal of plastic ureteral stents, which would reduce health care costs and improve the quality of life of patients.

The optimal combination of polymers and copolymers in the stent is determined by the requirements of their placement in the upper urinary tract [10,20]. The main current limitations to the development of biodegradable ureteral stents (BUS) are related to the control of the degradation rate and the size of the degradation fragments [10,20–23]. For this reason, a combination of polymers and copolymers with different degradation rates (always with degradation mainly by hydrolysis when placed in urine) are chosen to produce BUS. PGA is a synthetic, biodegradable, rapidly-degrading polymer, with glycolic acid as a metabolite. Poly-4-hydroxybutyrate is a long-term absorbable biomaterial, in which degradation occurs by hydrolysis and enzymatic pathways. 4-hydroxybutyric acid is a natural metabolite, which is primarily converted to carbon dioxide and water [24]. Finally, Glycomer$^{TM}$ 631 shows a degradation rate intermediate between these two options [25]. The application of polymers used in surgical sutures has been previously described. Due to their biocompatibility and ease of biodegradation, they may be a suitable material for drug delivery by techniques that include electrospinning, melt-extrusion, and coating [26].

For this reason, the two stent groups (BraidStent 1 and 2) use stents composed of a combination of two synthetic polymers with different degradation rates. This allows the degradation to be controlled as their biodegradation starts at different times, thus allowing the stent to fulfil its function as a scaffold while gradually reducing its mass, producing smaller fragments than if the stents were composed of a single degradable biomaterial. In protocol I, we found that both groups showed significant differences at T0 and T1 with respect to stent weight, but not stent thickness (Figures 4 and 9). We believe that the degradation characteristics of poly-4-hydroxybutyrate are the cause of the significantly slower biodegradation, although it should be noted that the degradation fragments were similar in both groups. An in vitro study by Barros et al. used artificial urine to assess a BUS from natural origin polymers (biopolymers) such as alginate as well as gellen and their blends with gelatine; it was found that the degradation duration ranged from 14 up to 60 days. This is in agreement with our results with synthetic polymers from protocol I, i.e., 52–77 days. Unlike synthetic polymers, biopolymers present better elasticity, biocompatibility, interface lubricity, and resistance to biofilm formation and encrustation [27].

One factor to consider in the assessment of new biodegradable stents in the urinary tract is that most synthetic polymers show a higher degradation rate at acidic pH than at physiological pH, which affects the ability to adjust the degradation time of the stent to the demands of patients [24,28]. Our experimental study demonstrates that, in all tested groups, excluding the BraidStent-SF-MMC1 group, the pH decreased significantly between baseline and T1, with a greater decrease in pH in the non-SF groups (pH < 6.0) (Figure 6). However, the range of pH changes found in our study, irrespective of the groups, remains within normal values for human beings. This is very relevant as significant alterations in urinary pH are associated with stent encrustation or crystal deposition that can lead to urinary lithiasis.

SF represents a new biomaterial, which, according to the scientific literature, can solve weaknesses related to stent coating, as it has the ability to carry drugs, control drug release kinetics, and improve the internal scaffold function [11]. The biocompatibility and biosafety of SF have been demonstrated previously, and the FDA has approved this polymer matrix for use in medical devices [12]. SF stent coating has been used for the delivery of various drugs, such as heparin, paclitaxel, clopidogrel, curcumin, 5-fluorouracil, sirolimus, emodin, theophylline, amoxicillin, and salicylic acid [11–13,29–31]. In the current experimental study, we added MMC to this series of drugs, which had not previously been evaluated as a drug to be included in an SF coating. Our results confirm that SF is useful for carrying and releasing MMC over a time period that could correspond to the postoperative application of chemotherapy (Table 1). Additionally, its biodegradation capacity allows the ureteral stent designed in our study to completely degrade. This degradation of the stent and its coating took place in a urinary environment, and it was achieved due to the layer-by-layer distribution with which it was designed. It is very important to control the changes in SF crystallization because the crystalline domains of SF are mainly formed by its beta-sheet structure, which contributes to MMC release. Our coating technique, by dipping the aqueous SF in absolute methanol, allowed the insolubilization process to increase the beta-sheet content of the regenerated SF. In our study, this allowed for the deposition of SF layer by layer, ensuring the encapsulation of MMC and its gradual and programmed release in the first 12 h, in order to achieve a suitable protocol for patients. In our methodology, to avoid losses in the encapsulation of MMC in the SF layers during methanol dipping, MMC was incorporated not only in the SF solution, but also in the methanol solution, at the same concentration. This represents an innovation with respect to the SF dip-coating techniques reported in the scientific literature.

An interesting attribute provided by SF is its ease of processability in different formats such as gels, membranes, coatings, and scaffolds [32]. This property was used in our study for the BraidStent-2 coating. Despite the statistically significant increase in thickness that SF and MMC caused, it is important to highlight that though thickness is below the size used in patients so it will not be an obstacle to clinical application [33].

Comparisons of this new design (BraidStent-SF-MMC1 or -2) with other biodegradable ureteral stents made of PGA and Glycomer[TM] 631 have been experimentally evaluated in the literature [21–23]. We found that, as with Braidstent groups 1 and 2 in the current study, coating with SF and MMC caused a significant reduction in the rate of ureteral stent degradation, from 6–7 weeks to 13–14 weeks (Figure 5) [21–23]. Despite significant differences in baseline thickness and weight between the BraidStent-SF, and the BraidStent-SF-MMC1- and -2 groups, no differences in overall degradation time were seen (Figures 4, 5 and 9). This may be due to the fact that, in both MMC formulations, complete release occurred within the first 12 h, a fact verified by HPLC-DAD. On the other hand, in a study assessing a very promising compound, such as Titanium dioxide nanoparticles to inhibit tumor development, which has shown in cell culture and in animal models similar results to current chemotherapy (Doxorubicin) but which reduces the side effects associated with conventional therapies. The researchers find, as in our study, that exposure at 6 and 12 h reach their maximal uptake and relate this to anti-tumor efficacy [34]. Regarding this MMC release rate, we found a huge difference between the two formulations, especially at 12 h, with a significantly higher concentration of MMC released by the BraidStent-SF-MMC2 group (Table 1). In this regard, unfortunately our study did not allow us to clarify whether the difference in the concentration of MMC released between the two groups depended exclusively on the different concentration of MMC added to the SF, or whether it was due to the use of pure MMC in the BraidStent-SF-MMC2 group. However, we did find greater difficulties in the manufacture of stents when MMC contained sodium chloride as an excipient compared to pure MMC, which did not cause precipitation of the dip solution. This is an aspect that will be assessed in future studies.

## 5. Conclusions

The present study demonstrated the feasibility of using MMC-loaded SF for the coating of biodegradable urinary stents. Coating with a fibroin matrix allowed for the controlled and programmed release of MMC but delayed the complete biodegradation of the ureteral stent. The addition of MMC to the coating of SF stents could be an attractive approach for intracavitary instillation in the upper urinary tract.

**Author Contributions:** Conceptualization, F.S. and S.D.A.-C.; methodology, F.S., S.D.A.-C. and J.E.d.l.C.; funding acquisition, F.S.; investigation, F.S., S.D.A.-C., A.B., J.A., J.E.d.l.C., J.P.C., Á.S. and F.M.S.M.; data curation, F.S.; writing–original draft preparation, F.S. and S.D.A.-C.; project administration, F.S.; writing–review and editing, F.S. All authors have read and agreed to the published version of the manuscript.

Polymers **2022**, 14, 3059

**Funding:** This research was funded by Instituto de Salud Carlos III (ISCIII, Spain) through the project PI16/01707 and PI20/01188. And project IB18107 funded by Consejería de Economía, Ciencia y Agenda Digital-Junta de Extremadura (Spain) and co-funded by the European Union. This study has been partially supported (80%) by the European Commission ERDF/FEDER Operational Programme 'Murcia' CCI No 2007ES161PO001 (Project No. 14–20/20).

**Institutional Review Board Statement:** Not applicable.

**Informed Consent Statement:** Not applicable.

**Data Availability Statement:** Not applicable.

**Acknowledgments:** Special gratefulness to Fernanda Carrizosa for her generous support during the writing of this manuscript, Luna Martínez-Plá for her technical support, and María Pérez for her BraidStent illustration.

**Conflicts of Interest:** The authors declare no conflict of interest. The funders had no role in the design of the study; in the collection, analyses, or interpretation of data; in the writing of the manuscript, or in the decision to publish the results.

## References

1.  Siegel, R.L.; Miller, K.D.; Fuchs, H.E.; Jemal, A. Cancer Statistics, 2021. *CA Cancer J. Clin.* **2021**, *71*, 7–33. [CrossRef] [PubMed]
2.  Benamran, D.; Seisen, T.; Naoum, E.; Vaessen, C.; Parra, J.; Mozer, P.; Shariat, S.F.; Rouprêt, M. Risk stratification for upper tract urinary carcinoma. *Transl. Androl. Urol.* **2020**, *9*, 1799–1808. [CrossRef] [PubMed]
3.  Rouprêt, M.; Babjuk, M.; Burger, M.; Capoun, O.; Cohen, D.; Compérat, E.M.; Cowan, N.C.; Dominguez-Escrig, J.L.; Gontero, P.; Hugh Mostafid, A.; et al. European association of urology guidelines on upper urinary tract urothelial carcinoma: 2020 Update. *Eur. Urol.* **2021**, *79*, 62–79. [CrossRef] [PubMed]
4.  Metcalfe, M.; Wagenheim, G.; Xiao, L.; Papadopoulos, J.; Navai, N.; Davis, J.W.; Karam, J.A.; Kamat, A.M.; Wood, C.G.; Dinney, C.P.; et al. Induction and Maintenance Adjuvant Mitomycin C Topical Therapy for Upper Tract Urothelial Carcinoma: Tolerability and Intermediate Term Outcomes. *J. Endourol.* **2017**, *31*, 946–953. [CrossRef]
5.  Gallioli, A.; Boissier, R.; Territo, A.; Vila Reyes, H.; Sanguedolce, F.; Gaya, J.M.; Regis, F.; Subiela, J.D.; Palou, J.; Breda, A. Adjuvant single-dose upper urinary tract instillation of mitomycin c after therapeutic ureteroscopy for upper tract urothelial carcinoma: A single-centre prospective non-randomized trial. *J. Endourol.* **2020**, *34*, 573–580. [CrossRef]
6.  Jung, H.; Giusti, G.; Fajkovic, H.; Herrmann, T.; Jones, R.; Straub, M.; Baard, J.; Osther, P.J.S.; Brehmer, M. Consultation on UTUC, Stockholm 2018: Aspects of treatment. *World J. Urol.* **2019**, *37*, 2279–2287. [CrossRef]
7.  Foerster, B.; D'Andrea, D.; Abufaraj, M.; Broenimann, S.; Karakiewicz, P.I.; Rouprêt, M.; Gontero, P.; Lerner, S.P.; Shariat, S.F.; Soria, F. Endocavitary treatment for upper tract urothelial carcinoma: A meta-analysis of the current literature. *Urol. Oncol.* **2019**, *37*, 430–436. [CrossRef]
8.  Porta, C.; Giannatempo, P.; Rizzo, M.; Lucarelli, G.; Ditonno, P.; Battaglia, M. An evaluation of UGN-101, a sustained-release hydrogel polymer-based formulation containing mitomycin-C, for the treatment of upper urothelial carcinomas. *Expert Opin. Pharmacother.* **2020**, *21*, 2199–2204. [CrossRef]
9.  Barros, A.A.; Browne, S.; Oliveira, C.; Lima, E.; Duarte, A.R.; Healy, K.E.; Reis, R.L. Drug-eluting biodegradable ureteral stent: New approach for urothelial tumors of upper urinary tract cancer. *Int. J. Pharm.* **2016**, *513*, 227–237. [CrossRef]
10. Soria, F.; de la Cruz, J.E.; Budia, A.; Serrano, A.; Galan-Llopis, J.A.; Sanchez-Margallo, F.M. Experimental Assessment of New Generation of Ureteral Stents: Biodegradable and Antireflux Properties. *J. Endourol.* **2020**, *34*, 359–365. [CrossRef]
11. Wang, X.; Zhang, X.; Castellot, J.; Herman, I.; Iafrati, M.; Kaplan, D.L. Controlled release from multilayer silk biomaterial coatings to modulate vascular cell responses. *Biomaterials* **2008**, *29*, 894–903. [CrossRef]
12. Xie, X.; Zheng, X.; Han, Z.; Chen, Y.; Zheng, Z.; Zheng, B.; He, X.; Wang, Y.; Kaplan, D.L.; Li, Y.; et al. A biodegradable stent with surface functionalization of combined-therapy drugs for colorectal cancer. *Adv. Healthc. Mater.* **2018**, *7*, e1801213. [CrossRef]
13. Xu, W.; Yagoshi, K.; Asakura, T.; Sasaki, M.; Niidome, T. Silk Fibroin as a Coating Polymer for Sirolimus-Eluting Magnesium Alloy Stents. *ACS App.l Bio Mater.* **2020**, *3*, 531–538. [CrossRef]
14. Wang, X.; Wenk, E.; Hu, X.; Castro, G.R.; Meinel, L.; Wang, X.; Li, C.; Merkle, H.; Kaplan, D.L. Silk coatings on PLGA and alginate microspheres for protein delivery. *Biomaterials* **2007**, *28*, 4161–4169. [CrossRef]
15. Sahoo, S.; Toh, S.L.; Goh, J.C. PLGA nanofiber-coated silk microfibrous scaffold for connective tissue engineering. *J. Biomed Mater. Res. B Appl. Biomater.* **2010**, *95*, 19–28. [CrossRef]
16. Aznar-Cervantes, S.D.; Pagan, A.; Monteagudo Santesteban, B.; Cenis, J.L. Effect of different cocoon stifling methods on the properties of silk fibroin biomaterials. *Sci. Rep.* **2019**, *9*, 6703. [CrossRef]
17. Rockwood, D.N.; Preda, R.C.; Yücel, T.; Wang, X.; Lovett, M.L.; Kaplan, D.L. Materials fabrication from Bombyx mori silk fibroin. *Nat. Protoc.* **2011**, *22*, 1612–1631. [CrossRef]

18. Matin, S.F.; Pierorazio, P.M.; Kleinmann, N.; Gore, J.L.; Shabsigh, A.; Hu, B.; Chamie, K.; Godoy, G.; Hubosky, S.G.; Rivera, M.; et al. Durability of Response to Primary Chemoablation of Low-Grade Upper Tract Urothelial Carcinoma Using UGN-101, a Mitomycin-Containing Reverse Thermal Gel: OLYMPUS Trial Final Report. *J. Urol.* **2022**, *207*, 779–788. [CrossRef]

19. Shan, H.; Cao, Z.; Chi, C.; Wang, J.; Wang, X.; Tian, J.; Yu, B. Advances in Drug Delivery via Biodegradable Ureteral Stent for the Treatment of Upper Tract Urothelial Carcinoma. *Front Pharmacol.* **2020**, *11*, 224. [CrossRef]

20. de la Cruz, J.E.; Soto, M.; Martínez-Plá, L.; Galán-Llopis, J.A.; Sánchez-Margallo, F.M.; Soria, F. Biodegradable ureteral stents: In vitro assessment of the degradation rates of braided synthetic polymers and copolymers. *Am. J. Clin. Exp. Urol.* **2022**, *10*, 1–12.

21. Soria, F.; de La Cruz, J.E.; Fernandez, T.; Budia, A.; Serrano, Á.; Sanchez-Margallo, F.M. Heparin coating in biodegradable ureteral stents does not decrease bacterial colonization-assessment in ureteral stricture endourological treatment in animal model. *Transl. Androl Urol.* **2021**, *10*, 1700–1710. [CrossRef]

22. Soria, F.; de La Cruz, J.E.; Caballero-Romeu, J.P.; Pamplona, M.; Pérez-Fentes, D.; Resel-Folskerma, L.; Sanchez-Margallo, F.M. Comparative assessment of biodegradable-antireflux heparine coated ureteral stent: Animal model study. *BMC Urol.* **2021**, *21*, 32. [CrossRef]

23. Soria, F.; de La Cruz, J.E.; Budia, A.; Cepeda, M.; Álvarez, S.; Serrano, Á.; Sanchez-Margallo, F.M. Iatrogenic Ureteral Injury Treatment with Biodegradable Antireflux Heparin-Coated Ureteral Stent-Animal Model Comparative Study. *J. Endourol.* **2021**, *35*, 1244–1249. [CrossRef]

24. Odermatt, E.K.; Funk, L.; Bargon, R.; Martin, D.P.; Rizk, S.; Williams, S.F. MonoMax Suture: A New Long-Term Absorbable Monofilament Suture Made from Poly-4-Hydroxybutyrate. *J. Polym. Sci.* **2012**, *2012*, 216137. [CrossRef]

25. Molea, G.; Schonauer, F.; Bifulco, G.; D'Angelo, D. Comparative study on biocompatibility and absorption times of three absorbable monofilament suture materials (Polydioxanone, Poliglecaprone 25, Glycomer 631). *Br. J. Plast Surg.* **2000**, *53*, 137–141. [CrossRef]

26. Deng, X.; Qasim, M.; Ali, A. Engineering and polymeric composition of drug-eluting suture: A review. *J. Biomed Mater. Res. A* **2021**, *109*, 2065–2081. [CrossRef]

27. Barros, A.A.; Barros, A.A.; Rita, A.; Duarte, C.; Pires, R.A.; Sampaio-Marques, B.; Ludovico, P.; Lima, E.; Mano, J.F.; Reis, R.L. Bioresorbable ureteral stents from natural origin polymers. *J. Biomed Mater Res. B Appl. Biomater.* **2015**, *103*, 608–617. [CrossRef]

28. Chu, C.C. A comparison of the effect of pH on the biodegradation of two synthetic absorbable sutures. *Ann. Surg.* **1982**, *195*, 55–59. [CrossRef]

29. Cheema, S.K.; Gobin, A.S.; Rhea, R.; Lopez-Berestein, G.; Newman, R.A.; Mathur, A.B. Silk fibroin mediated delivery of liposomal emodin to breast cancer cells. *Int. J. Pharm.* **2007**, *341*, 221–229. [CrossRef]

30. Bayraktar, O.; Malay, O.; Ozgarip, Y.; Batigün, A. Silk fibroin as a novel coating material for controlled release of theophylline. *Eur. J. Pharm. Biopharm.* **2005**, *60*, 373–381. [CrossRef]

31. Rujiravanit, R.; Kruaykitanon, S.; Jamieson, A.M.; Tokura, S. Preparation of crosslinked chitosan/silk fibroin blend films for drug delivery system. *Macromol. Biosci.* **2003**, *3*, 604–611. [CrossRef]

32. Wang, X.; Hu, X.; Daley, A.; Rabotyagova, O.; Cebe, P.; Kaplan, D.L. Nanolayer biomaterial coatings of silk fibroin for controlled release. *J. Control Release* **2007**, *121*, 190–199. [CrossRef] [PubMed]

33. Diatmika, A.A.N.O.; Djojodimedjo, T.; Kloping, Y.P.; Hidayatullah, F.; Soebadi, M.A. Comparison of ureteral stent diameters on ureteral stent-related symptoms: A systematic review and meta-analysis. *Turk J. Urol.* **2022**, *48*, 30–40. [CrossRef] [PubMed]

34. Iqbal, H.; Razzaq, A.; Uzair, B.; Ul Ain, N.; Sajjad, S.; Althobaiti, N.A.; Albalawi, A.E.; Menaa, B.; Haroon, M.; Khan, M.; et al. Breast Cancer Inhibition by Biosynthesized Titanium Dioxide Nanoparticles Is Comparable to Free Doxorubicin but Appeared Safer in BALB/c Mice. *Materials* **2021**, *14*, 3155. [CrossRef]

Article

# Production and Characterization of Active Bacterial Cellulose Films Obtained from the Fermentation of Wine Bagasse and Discarded Potatoes by *Komagateibacter xylinus*

Patricia Cazón, Gema Puertas and Manuel Vázquez *

Department of Analytical Chemistry, Faculty of Veterinary, University of Santiago de Compostela, 27002 Lugo, Spain
* Correspondence: manuel.vazquez@usc.es

**Abstract:** Potato waste, such as peels, broken or spoiled potatoes and grape bagasse residues from the winery industry, can be used for the biotechnological production of high-value products. In this study, green, sustainable and highly productive technology was developed for the production of antioxidant bacterial cellulose (BC). The aim of this work was to evaluate the feasibility of a low-cost culture medium based on wine bagasse and potato waste to synthesize BC. Results show that the production of BC by *Komagateibacter xylinus* in the GP culture medium was five-fold higher than that in the control culture medium, reaching 4.0 g/L BC in 6 days. The compounds of the GP culture medium improved BC production yield. The mechanical, permeability, swelling capacity, antioxidant capacity and optical properties of the BC films from the GP medium were determined. The values obtained for the tensile and puncture properties were 22.77 MPa for tensile strength, 1.65% for elongation at break, 910.46 MPa for Young's modulus, 159.31 g for burst strength and 0.70 mm for distance to burst. The obtained films showed lower permeability values ($3.40 \times 10^{-12}$ g/m·s·Pa) than those of other polysaccharide-based films. The BC samples showed an outstanding antioxidant capacity (0.31–1.32 mg GAE/g dried film for total phenolic content, %DPPH$^\bullet$ 57.24–78.00% and %ABTS$^{\bullet+}$ 89.49–86.94%) and excellent UV-barrier capacity with a transmittance range of 0.02–0.38%. Therefore, a new process for the production of BC films with antioxidant properties was successfully developed.

**Keywords:** bacterial cellulose; agricultural waste; *Komagateibacter xylinus*; active film; antioxidant properties; ABTS; DPPH; TPC

**Citation:** Cazón, P.; Puertas, G.; Vázquez, M. Production and Characterization of Active Bacterial Cellulose Films Obtained from the Fermentation of Wine Bagasse and Discarded Potatoes by *Komagateibacter xylinus*. *Polymers* **2022**, *14*, 0. https://doi.org/

Academic Editor: Piotr Bulak

Received: 29 October 2022
Accepted: 22 November 2022
Published: 29 November 2022

**Publisher's Note:** MDPI stays neutral with regard to jurisdictional claims in published maps and institutional affiliations.

## 1. Introduction

In 2019, the Food and Agriculture Organization of the United Nations estimated that approximately 14% of the world's food, valued at 400 billion USD, is lost on an annual basis between harvesting and the retail market (https://www.fao.org, (accessed on 24 October 2022)). Food loss and waste has indeed become an issue of great public concern. The 2030 Agenda for Sustainable Development reflects the increased global awareness of the problem. Target 12.3 of the Sustainable Development Goals calls for halving global per capita food waste at retail and consumer levels by 2030, as well as reducing food losses along production and supply chains (https://www.fao.org, (accessed on 24 October 2022)). Food loss and waste streams often contain valuable components that should be considered to transform the current linear supply chain into a circular model, following the foundations of a circular economy. In addition, these resources can help address the major global challenge of transitioning from a fossil-based to a bio-based economy, including energy, chemicals and materials. Food loss and waste have been assessed mainly for biofuel production and energy recovery and, to a lesser extent, for synthesizing bio-based materials. Due to the need to find biodegradable materials to help minimize the consumption of non-biodegradable packaging with a higher environmental impact, there has been an increased interest in evaluating these wastes as feedstock for the production of biomaterials [1]. Often, food loss and waste can be used directly without separation or purification processes as a source of nutrients for the fermentation of microorganisms that produce biopolymers such as polyhydroxyalkanoates or bacterial cellulose (BC), reducing production costs [2,3].

BC, like vegetable cellulose, is a linear polysaccharide of covalently linked glucose residues between carbon 1 and 4 (β1–4). BC has several advantages over vegetable cellulose: (1) BC is synthesized in a pure form, free of other vegetable impurities; (2) it does not require costly extraction and purification processes or the use of environmentally hazardous chemicals; (3) it has a higher degree of crystallinity and polymerization; (4) it has superior tensile properties; and (5) it has a higher specific area and swelling capacity due to its network structure [4]. BC is presented as a promising biodegradable material with applications in multiple fields, such as biomedicine [5], pharmaceuticals, cosmetics and the food industry [6,7], among others. However, the large-scale industrialization and commercialization of BC remains a challenge due to high fermentation costs, low productivity and expensive culture media. In order to reduce the costs of culture media, research has been intensifying to evaluate low-cost biomass waste substrates as a nutrient source for BC production [8]. Among waste generated in greater volumes and due to their chemical compositions, waste from wineries and the potato industry can be excellent candidates for BC production.

The International Organization of Vine and Wine estimated that, in 2018, the global production of grapes reached 77.8 Mt, being 57% of the harvest destined for wine production (https://www.oiv.int, (accessed on 24 October 2022)). During the production of red wine, after fermentation and pressing, a solid residue is obtained consisting of stalks, skins and seeds, namely grape pomace or grape marc. Grape marc represents approximately 20% of the total processed grapes and is rich in polyphenols, anthocyanins, flavonoids, essential oils and polysaccharides, with potential revaluation applications [9]. The rich composition of wine residues in these components allows value-added products to be obtained through biotechnological processes [9]. Previous studies have evaluated the use of grapes in several versions (juice, must and unpressed residue) as a cheap carbohydrate source to produce BC. Mostly unpressed residues or grape extract with a high content of free monosaccharides are used as alternative carbon sources in Hestrin–Schramm media or in alternative media enriched with corn steep liquor as a nitrogen source [10–13]. However, most waste generated in wineries is grape marc, which, after fermentation and pressing, has a limited content of monosaccharides available for fermentation. On the other hand, in many studies, bagasse components (seeds, stalk, pulp or skin) are separated and subjected to separation processes in order to use mainly the components with the highest monosaccharide content [10,12]. Additional operating units increase the cost of production. In this work, the whole residue is evaluated, avoiding the separation of components, as a source of monosaccharides and nutrients that improve the yield of BC production.

The potato (*Solanum tuberosum*), which is the fourth leading crop after rice, wheat and maize, plays an important role in the human diet worldwide. The worldwide production of potatoes is constantly increasing and currently amounts to around 380 Mt. The share of potatoes in 2011 designated as a raw material for the industry was 130 Mt [14,15]. The main compound of potato is starch. It is composed of two polysaccharides: the linear molecule amylose, which consists of glucose polymers connected predominantly by α-1,4 bonds, and amylopectin, the main macromolecular component of starch and responsible for the structure of starch granules [16]. The contribution of potato starch to total starch production in the EU is 15–20% and 3–4% worldwide. In recent years, potato starch production in Europe has been approximately 1.7 million tons per year [17]. Due to their content of starch, cellulose, hemicelluloses and fermentable sugars, potato residues (non-commercial potatoes, potato peelings, potatoes with high content of glucoalkaloids α-solanine, and α-chaconine) are a good substrate for biotechnological processes [16]. A previous study evaluated the use of potato peel waste [18] for the low-cost production of BC. In that case, the culture medium contains only sugars from the acid hydrolysate of the potato peels, without being enriching with other components or hydrolysates of other residues with a high content of micronutrients or antioxidants that can increase the yield of BC production or that can obtain functionalized films with antioxidant properties.

The acid hydrolysis treatment of potato and wine bagasse residues is a simple and economical option to break down cellulose and hemicellulose structures and to increase the content of fermentable monosaccharides.

The use of the potential high content of hydrolysable polysaccharides to obtain fermentable sugars, as well as the possible presence of micronutrients, mainly in the remains of grape bagasse, could turn these agri-food wastes into an excellent source of substrates to obtain enriched culture media for economically more profitable biotechnological processes, such as the production of BC. Moreover, the presence of antioxidant components in the bagasse could allow in situ modifications of the BC films, retaining these antioxidant components within the BC polymeric matrix, leading to films with antioxidant properties after fermentation, purification and drying processes.

The aim of this work was to evaluate the feasibility of using a low-cost culture medium based on Garnacha bagasse and potato waste to synthesize BC. The transfer of the antioxidants of the bagasse extract to the polymeric matrix was evaluated to obtain active BC films with antioxidant properties. Moreover, the mechanical, permeability, swelling, antioxidant and optical properties of the obtained films were measured.

## 2. Materials and Methods

### 2.1. Chemicals and Standards

For the synthesis of BC, a pure freeze-dried culture of *Komagateibacter xylinus* was obtained from the Spanish Type Culture Collection "Colección Española de Cultivos Tipo" (CECT) (Paterna, Spain). The control medium was prepared with D(+)-glucose monohydrate (99%) provided by Acros Organics (Geel, Belgium) and yeast extract supplied by Scharlau (Barcelona, Spain). The natural culture medium that was evaluated was based on bagasse extracts and potato hydrolysates obtained under acid hydrolysis conditions using sulfuric acid (95–97%), and sodium carbonate purchased from Scharlau Microbiology (Barcelona, Spain) was used to neutralize the media.

Grape pomace from *Vitis vinifera* L. Garnacha Tintorera, which contains skins, pulp, seeds and stems after the pressing process in winemaking, was kindly provided by a local winery (42°33'58.2" N 7°40'37.4" W) in the Ribeira Sacra region (Lugo, Spain). Tuber samples of potatoes (*Solanum tuberosum*) were generously supplied by a local company called Pitita's Farm (Dozón, Spain), harvested at the following geocoordinates: 42°36'14.0" N 8°01'24.6"W.

The synthesized BC samples were purified using sodium hydroxide (98%) and were conditioned using anhydrous sodium bromide (99%) or silica gel from Acros Organics (Geel, Belgium). All reagents and solvents used for HPLC were of analytical HPLC grade. Standards of gluconic acid, D(+)-glucose monohydrate (99%), D-xylose (99%), L(+)-arabinose (99%), 5-(hydroxymethyl)furfural (98%) and furfural (99%) standards were purchased from Acros Organics (Geel, Belgium). Folin–Ciocalteu reagent from Panreac (Barcelona, Spain), 1,1-diphenyl-2-picrylhydrazyl radicals (DPPH•) and 2,2'-azino-bis(3-ethylbenzothiazoline-6-sulfonic acid) (ABTS•+) free radical reagents purchased from Alfa Aesar (Haverhill, MA, USA) and methanol and ethanol provided by Scharlau Microbiology (Barcelona, Spain) were used for antioxidant assays.

### 2.2. Culture Media for Bacterial Cellulose Synthesis

Grape pomace, which contained skins, pulp, seeds and stems, was dried using hot air-drying at 50 °C for 24 h. The dried grape pomace was ground into powder and was sieved by a 0.5 mm mesh sieve. The homogeneous sample with a size lower than 0.5 mm was stored in an airtight container at room temperature until use. The same process was followed to dry and homogenize the potato samples after washing and slicing. Aliquots of the homogenized raw material were analyzed for moisture determination by drying a known amount of sample at 105 °C to a constant weight. The moisture content test was carried out in triplicate.

The sulfuric acid hydrolysis was performed in an autoclave capable of controlling temperature (±1 °C). Samples with a bagasse powder/potato powder ratio of 50:50 ($w/w$) were prepared in 0.5 L bottles by adding a 2% sulfuric acid solution to a final solid/liquid ratio of 1:10. In order to allow all the matter to be wetted and to obtain good supernatant recovery, the mixture was kept under vigorous stirring for 1 h before the thermal process. Then, the reaction mixture was submitted at 125 °C for 1 h. When the heat treatment was completed and the mixture was cooled to room temperature, the hydrolysate liquors were neutralized with $CaCO_3$ until reaching a pH of 6. The resulting solution was filtered through filter paper (10 μm porosity) to remove un-dissolved material, and the $CaSO_4$ was precipitated from the supernatant. The natural grape pomace–potato (GP) culture media were prepared with the hydrolysates and were supplemented with 10 g of yeast extract/L.

The selected synthetic culture media were recommended by the Spanish Type Culture Collection (CECT), equaling the glucose concentration of that obtained in the natural medium studied and with 10 g of yeast extract/L.

*2.3. Bacterial Cellulose Fermentation*

Pre-inoculum was prepared from a starter stock of *K. xylinus*. For this purpose, 100 mL of GP culture medium and 100 mL of control synthetic culture medium were prepared in 250 mL Erlenmeyer flasks and were sterilized at 121 °C for 15 min. The sterile media were inoculated with the *K. xylinus* starter culture and were incubated under static conditions for 2 days at 30 °C.

For the BC production study, Petri dishes were prepared with 25 mL of culture medium (GP or control) and were inoculated with 10% of the volume of the corresponding pre-inoculum. The dishes were incubated under static conditions for a maximum of 10 days at 30 °C. Each day, 3 dishes were removed to analyze the composition of the culture medium and the BC film formed in the air–liquid medium. All experiments were carried out in triplicate, and the means and standard deviations are given. The parameters that were measured to analyze the production of BC in the media under study were as follows:

- The pH was measured using pHmeter pH 210 (Hanna Instruments, Woonsocket, RI, USA).
- The drained weight (g) was determined from the weight of the samples of BC, from which excess liquid was removed by draining the BC for 1 h.
- The dry weight (g) of the BC was determined from samples dried at 105 °C for 24 h.
- The swelling capacity (%S) of the polysaccharides was calculated from the difference in the drained weight and the dry weight of the polymer matrix.
- BC concentration (g/L) was calculated by considering the amount of BC synthesized (dry weight, g) per unit of volume of the culture medium in each Petri dish (25 mL).
- BC productivity (g/L·h), BC yield and BC efficiency were calculated using Equations (1)–(3) [19,20]:

$$BC\ productivity\ (g/L \cdot h) = X/V \cdot t \tag{1}$$

$$BC\ yield = (X - X_0)/S_0 - S \tag{2}$$

$$BC\ efficiency = (X - X_0)/S_0 \tag{3}$$

where X is the biomass concentration under steady-state conditions (g/L), V is the reaction volume (L), t is the time of reaction (d), $X_0$ is the initial biomass concentration (g/L), S is the glucose concentration (g/L) and $S_0$ is the initial glucose concentration (g/L).

*2.4. Analytical Methods*

Glucose, gluconic acid, cellobiose, xylose, arabinose, hydroxymethyl furfural and the furfural content of the culture media were determined by HPLC using a Rezex RHM

(Phenomenex, Torrance, CA, USA) column with isocratic elution (flow rate of 0.400 mL/min and mobile phase of 0.025 M $H_2SO_4$), a column oven set at 45 °C and a refractive index detector (RI) (LC 2000 plus, Jasco, Tokio, Japan). The RI detector was used for the analysis of glucose, cellobiose, xylose and arabinose. Gluconic acid, furfural and HMF were detected spectrophotometrically with DAD. Gluconic acid was detected at 220 nm, and furfural and HMF were detected at 275 nm [21–23]. The analysis was performed in triplicate by taking 3 Petri dishes each day until the end of the incubation (10 days).

### 2.5. Bacterial Cellulose Films Processing

Based on the optimal conditions of the minimum time required to achieve the highest BC production observed in the fermentation study in synthetic and alternative media, *K. xylinus* was fermented under static conditions at 30 °C. After the fermentation time, the formation of a cellulose film floating on the surface of the culture medium was observed. The pellicles were separated from the culture media and were washed with water to remove the culture medium remains. The homogeneous BC films were treated with a solution of NaOH 1% (*w/v*) at 90 °C for 1 h to remove bacterial cells. Later, films were subjected to several washing cycles with distilled water until a pH of 7 of the wash water was reached. The wet BC pellicles were placed between a support designed to dry the samples and to prevent them from wrinkling or shrinking. The films were dried for 48 h at room temperature. Finally, the films were cut into a specific size, and the thickness of each sample was measured at 5 random points using a thickness meter ET115S (Etari GmbH, Stuttgart, Germany). The samples were stored in desiccators with a saturated anhydrous sodium bromide solution or silica gel for 5 days, which was enough time to ensure that the samples reached an equilibrium moisture content.

### 2.6. Characterization of the Antioxidant Bacterial Cellulose Samples as Films

The antioxidant capacity of the samples was determined by the ABTS$^{\bullet+}$ (2,2'-azino-bis (3-ethylbenzothiazoline-6-sulfonic acid) and DPPH$^{\bullet}$ (1,1-diphenyl-2-picrylhydrazyl radicals) free radical scavenging assays, and the total phenolic content (TPC) was determined by the Folin–Ciocalteu method, as described elsewhere [24,25]. The TPC was expressed as mg of gallic acid equivalents (GAE) per 1 g of dried BC sample determined by calibration with gallic acid. The obtained calibration equation was TPC (mg ac gallic/g sample) = 1010 $Abs_{765nm} - 85.311$, and the determination coefficient was $R^2 = 0.9993$.

The equilibrium moisture content (%W) and the swelling capacity (%S) were estimated by gravimetric methods, following the previously reported method [26,27]. Water vapor permeability (WVP) was measured following the ASTM Standard Test Method E96 (https://www.astm.org/Standards/E96.htm, (accessed on 20 October 2022)). The WVP of the samples was determined at 30 °C and 50% relative humidity for 4 h, which was enough time to reach a dynamic equilibrium in the water flux. The equilibrium moisture content and WVP tests were carried out in triplicate.

The tensile strength (TS, MPa), percentage of elongation at break (%E, %), Young's Modulus (YM, MPa), burst strength (BS, g) and distance to burst (DB, mm) were measured with a texturometer (TA-XT plus, Stable Micro System, UK). The tensile test was performed using the standard method D-882 (ASTM), and samples had a size of 15 × 100 and were preconditioned for 5 days at room temperature and 57% relative humidity. The samples were clamped between the grips of the texturometer with an initial gap of 40 mm and a test speed of 1 mm/s. Seven samples were tested to calculate the mean value of each parameter [28,29]. A film holder (HDP/FSR) attached to the texturometer was used for the puncture test. Sample squares with 30 mm sides were conditioned at 57% relative humidity and room temperature for 5 days. The test was carried out in triplicate with a cylindrical probe (d = 3 mm) at a constant crosshead speed of 1 mm/s until rupture.

The UV barrier, transparency, opacity and color of the samples were measured with absorbance values in the UV-Vis light range (190–800 nm) using a spectrophotometer, V-670 (Jasco Inc., Tokyo, Japan), as described elsewhere [29]. The transparency and opacity values of the samples were estimated using their transmittance and absorbance values at 500 nm and 600 nm wavelengths. Spectra Manager software (Jasco Inc., Tokyo, Japan) was used to determine the CIE L*a*b* coordinates.

### 2.7. Statistical Analysis

The obtained results were statistically analyzed by a one-way analysis of variance (ANOVA) by employing Microsoft Excel® software. Differences between pairs of means were assessed based on confidence intervals using the Tukey's post hoc test. The least significant difference was $p < 0.05$.

### 3. Results and Discussion

The moisture contents of bagasse powder and potato powder dried with the air-dryer were $(2.38 \pm 0.06)\%$ and $(7.72 \pm 0.06)\%$, respectively. These values were considered when preparing the hydrolysis mixtures.

A minimum of 36 Petri dishes were prepared for the tests analyzed with the GP culture medium, and the same number of Petri dishes was prepared for the control culture medium. Every 24 h, from the start time of the incubation of *K. xylinus*, three plates were removed to evaluate and measure the different parameters that determined the changes in the composition of the culture medium and the growth of the BC. Figures 1 and 2 show the composition changes in the main analyzed components of the GP medium and the control culture medium throughout the incubation period. Tables 1 and 2 show the values obtained for the different parameters related to the production of BC in the GP and the control medium, respectively.

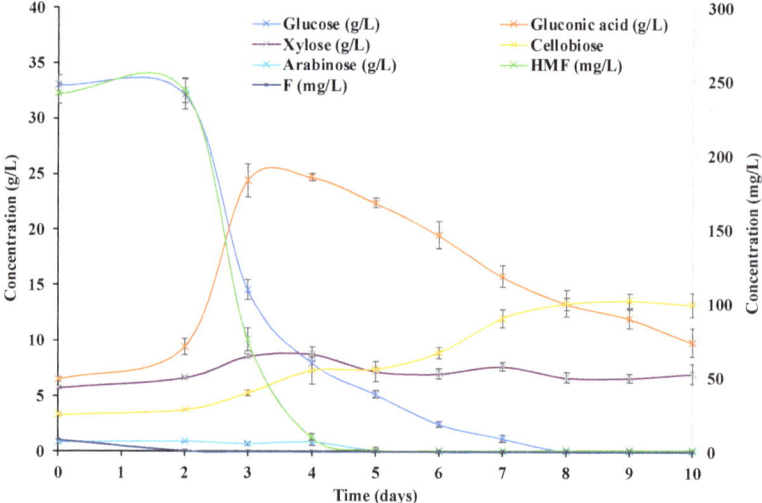

**Figure 1.** Cellobiose, gluconic acid, glucose, xylose, arabinose, 5-(hydroxymethyl)-2-furaldehyde (HMF) and furfural (F) concentration (g/L; mg/L) of the bagasse–potato culture medium during the incubation time. Values are shown as means and standard deviations.

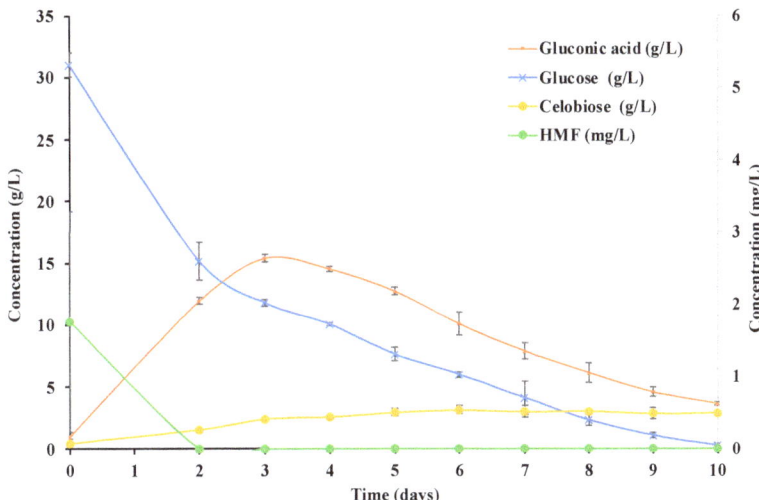

**Figure 2.** Cellobiose, gluconic acid, glucose and 5-(hydroxymethyl)-2-furaldehyde (HMF) concentration (g/L; mg/L) of the control culture medium during the incubation time. Values are shown as means and standards deviation.

**Table 1.** Parameters analyzed at each incubation time in bagasse–potato culture medium: pH, bacterial cellulose (BC) drained weight (g), BC dried weight (g), BC concentration (g/L), BC swelling capacity (%S) and gluconic acid concentration (g/L).

| Time | pH | Drained Weight | Dried Weight | BC | BC %S | Gluconic Acid |
|---|---|---|---|---|---|---|
| Day | | g | g | g/L | % | g/L |
| 0 | 5.20 ± 0.01 [a] | - | - | - | - | 6.49 ± 0.19 [a] |
| 2 | 4.68 ± 0.11 [b] | 0.48 ± 0.15 [a] | 0.01 ± 1.18 × 10$^{-3}$ [a] | 0.23 ± 0.05 [a] | 8611 ± 3001 [abe] | 9.43 ± 0.74 [b] |
| 3 | 3.57 ± 0.04 [cd] | 2.15 ± 0.10 [b] | 0.02 ± 2.73 × 10$^{-3}$ [a] | 0.67 ± 0.11 [a] | 12,935 ± 2551 [a] | 24.42 ± 1.47 [c] |
| 4 | 3.46 ± 0.01 [c] | 3.02 ± 0.14 [bc] | 0.041 ± 4.01 × 10$^{-3}$ [a] | 1.64 ± 0.16 [ae] | 7318 ± 745 [bc] | 24.72 ± 0.33 [c] |
| 5 | 3.50 ± 0.01 [cd] | 3.23 ± 0.50 [bc] | 0.09 ± 0.02 [b] | 3.66 ± 0.75 [h] | 3463 ± 419 [c] | 22.39 ± 0.43 [c] |
| 6 | 3.55 ± 0.01 [cdc] | 3.71 ± 0.19 [cc] | 0.10 ± 0.01 [b] | 4.00 ± 0.49 [b] | 3640 ± 393 [c] | 19.47 ± 1.22 [d] |
| 7 | 3.58 ± 0.01 [bde] | 4.69 ± 0.56 [ef] | 0.08 ± 0.01 [b] | 3.10 ± 0.33 [bce] | 5952 ± 217 [ec] | 15.79 ± 0.96 [e] |
| 8 | 3.64 ± 0.05 [bef] | 4.80 ± 0.41 [ef] | 0.11 ± 0.03 [b] | 4.32 ± 1.07 [bc] | 4615 ± 1646 [ec] | 13.36 ± 1.19 [f] |
| 9 | 3.70 ± 0.05 [bf] | 5.18 ± 0.20 [f] | 0.10 ± 0.01 [b] | 4.14 ± 0.48 [b] | 4963 ± 722 [ec] | 12.00 ± 0.82 [fg] |
| 10 | 3.80 ± 0.05 | 5.28 ± 0.76 [f] | 0.09 ± 0.01 [b] | 3.71 ± 0.30 [b] | 5658 ± 1276 [ec] | 9.86 ± 1.32 [g] |

Values are expressed as mean ± standard deviation (SD). Different letters in the same column indicate significant differences ($p > 0.05$).

**Table 2.** Parameters analyzed at each incubation time in control culture medium: pH, bacterial cellulose (BC) drained weight (g), BC dried weight (g), BC concentration (g/L), BC swelling capacity (%S) and gluconic acid concentration (g/L).

| Time | pH | Drained Weight | Dried Weight | BC Concentration | BC %S | Gluconic Acid |
|---|---|---|---|---|---|---|
| Day | | g | g | g/L | % | g/L |
| 0 | 5.26 ± 0.01 [a] | - | - | - | - | 0.98 ± 0.15 [a] |
| 2 | 3.25 ± 0.03 [bd] | 2.47 ± 0.48 [a] | 0.009 ± 0.001 | 0.35 ± 0.02 [a] | 28,169 ± 4038 [a] | 11.95 ± 0.28 [b] |
| 3 | 3.17 ± 0.01 [c] | 2.71 ± 0.20 [a] | 0.010 ± 3.06 × 10$^{-4}$ | 0.41 ± 0.01 [a] | 26,592 ± 1306 [a] | 15.40 ± 0.33 [c] |
| 4 | 3.17 ± 0.01 [c] | 2.61 ± 0.40 [a] | 0.010 ± 0.002 | 0.41 ± 0.07 [a] | 25,471 ± 1249 [a] | 14.53 ± 0.20 [c] |
| 5 | 3.15 ± 0.02 [c] | 2.29 ± 0.50 [a] | 0.011 ± 0.001 | 0.43 ± 0.05 [a] | 20,948 ± 4246 [abc] | 12.72 ± 0.31 [b] |

Table 2. *Cont.*

| Time | pH | Drained Weight | Dried Weight | BC Concentration | BC %S | Gluconic Acid |
|------|-----|----------------|--------------|------------------|-------|---------------|
| Day | | g | g | g/L | % | g/L |
| 6 | 3.18 ± 0.02 [bc] | 2.61 ± 0.21 [b] | 0.020 ± 0.003 | 0.78 ± 0.13 [b] | 13,352 ± 1487 [bc] | 10.08 ± 0.93 [d] |
| 7 | 3.27 ± 0.01 [d] | 2.71 ± 0.55 [a] | 0.012 ± 0.004 | 047 ± 0.14 [a] | 22,496 ± 2653 [abc] | 7.88 ± 0.67 [e] |
| 8 | 3.27 ± 0.03 [d] | 2.65 ± 0.04 [ab] | 0.014 ± 0.004 | 0.54 ± 0.17 [a] | 20,787 ± 5874 [ac] | 6.11 ± 0.80 [f] |
| 9 | 3.35 ± 0.06 [e] | 2.39 ± 0.87 [a] | 0.011 ± 0.0031 | 0.44 ± 0.10 [a] | 21,205 ± 2961 [a] | 4.57 ± 0.36 [g] |
| 10 | 3.42 ± 0.02 [f] | 2.09 ± 0.44 [ab] | 0.013 ± 0.001 | 0.52 ± 0.02 [a] | 16,100 ± 3956 [c] | 3.61 ± 0.14 [h] |

Values are expressed as mean ± standard deviation (SD). Different letters in the same column indicate significant differences ($p > 0.05$).

### 3.1. Bacterial Cellulose Production

The production of BC was tested on pure media of Garnacha bagasse hydrolysates. However, no BC production was observed in pure Garnacha bagasse media. The HPLC analysis of the carbon sources of Garnacha hydrolysate revealed a low concentration of sugars such as glucose (<5 g/L), xylose (<5 g/L) or arabinose (<2 g/L). Low concentrations of the carbon sources were likely not enough to form homogeneous BC films on the plate surfaces. The content of monosaccharides in the bagasse was low because the many sugars were previously consumed by yeasts during alcoholic fermentation in the wine production process. The used bagasse was previously drained and pressed, and therefore, almost all the juice with free sugars were extracted, being the final residue of the winemaking process. In other studies, a concentration of monosaccharides up to 17.6% or 13.64% was observed in grape extracts obtained directly from grape juice extract enriched with sugar cane (5%) [13] or from grapes under enzymatic hydrolysis treatment [2]. Furthermore, the content of monosaccharides and total sugars in grapes depends on the type of variety and the part of the grape analyzed [30].

Despite the low content of free monosaccharides, this extract contains a high content of phenolic compounds, antioxidant components or micronutrients as vitamin C that are of great interest to be included in active biodegradable films [2,31,32]. Moreover, these components can enrich the culture medium and enhance the synthesis of polysaccharides, such as BC [2,33]. The evaluated Garnacha hydrolysate showed a TPC value of 79.3 ± 6.7 mg GAE/L of extract, the %DPPH$^\bullet$ was (37.4 ± 6.9)% and the %ABTS$^{\bullet+}$ was (30.0 ± 6.2)%, indicating valuable content of phenolic compounds and other antioxidants in the bagasse hydrolysates.

The enrichment of Garnacha bagasse hydrolysate with non-commercial potato waste hydrolysate may increase the monosaccharide content (mainly glucose) of the culture medium to enhance the growth of *K. xylinus* and the BC production. A previous work evaluated the acid hydrolysis with sulfuric acid of non-commercial potatoes to obtain glucose for fermentative purposes [16]. The data showed a glucose concentration of 73.08 ± 0.09 g/L after a hydrolysis treatment at 130 °C for 60 min with a 2% sulfuric acid solution and a solid/liquid ratio of 1:10 [16].

Based on these results and with the goal of improving the monosaccharide content of the culture medium, culture media enriched with potato residues were prepared with a bagasse:potato ratio of 50:50 (*w/w*). The selected hydrolysis conditions were like the previous work, at 125 °C for 60 min with a sulfuric acid concentration of 2%. A slightly lower hydrolysis temperature was chosen to minimize the formation of undesirable components in the hydrolysate, such as HMF or furfural.

Figures 1 and 2 show the content of the main monosaccharides and disaccharides in the fermentation of *K. xylinus* in the GP and the control culture media. Starting from a GP hydrolysate of 50:50 (*w/w*), a medium with a glucose concentration of 33.0 g/L, an arabinose concentration of 0.9 g/L and a xylose concentration of 5.7 g/L was obtained. The glucose content values are in agreement with those previously obtained for pure hydrolysates of potato waste [16] and are higher than those reported in a previous study of acid hydrolyzed potato peelings [18]. The glucose in the medium came mainly from the hydrolysis of potato starch and in a small proportion from bagasse cellulose [16,30]. The small number of other monosaccharides detected in the hydrolysates was due to the presence of pectic polysaccharides and the hemicellulosic polysaccharides of grape bagasse [30]. For this reason, these monosaccharides were not detected in the control culture medium (Figure 2).

Glucose consumption can be related to both bacterial growth and BC production since it is the main carbon source in the medium. Comparing the glucose consumption of the GP medium with the control medium, a sharp decrease in glucose content in the GP medium was observed on the 3rd day, and this decrease took place in the control on the 2nd day. This could be due to the fact that the bacteria needed more time to adapt themselves to the new medium, but once the bacteria adapted, the glucose consumption was higher compared to the control. Thus, in the GP medium, all the glucose was consumed in less than 8 days, and in the control, it was on day 10. The presence of micronutrients in the GP medium led to an increase in the activity of the bacteria, with an increased rate of consumption of the main carbon source (glucose).

BC production in the GP medium was much higher than BC production in the control medium (Tables 1 and 2). After 6 days, a BC concentration of 0.78 g/L was reached in the control medium, obtaining an average of 0.02 g of dry BC per 25 mL of medium. However, this production was five times higher in the GP medium, reaching a BC concentration of 4.0 g/L and an average of 0.10 g of dry BC per 25 mL of GP medium. Note that, in the control medium on day 2, unlike the GP medium, BC production slowed down without increasing significantly during the next 8 days (Tables 1 and 2). The BC concentration values achieved in the present study were higher than those observed in previous work using grape pomace extract or potato residues. Grape pomace as a carbon source and corn steep liquor as the main nitrogen source needs 18 days of incubation to reach a concentration of over 4 g/L [10]. In the case of HS medium with white grape bagasse replacing the carbon source, incubated for 14 days at 28 °C, the BC concentration was 1.2 g/L [12]. In culture media containing pure xylose, glucose: galactose (1:1), lactose or glycerol, the BC production was 0.26, 2.60, 0.31 and 2.07 g/L, respectively [11]. The application of acid hydrolysates of potato peel waste for BC fermentation resulted in BC production of 1.21–2.61 g/L after 96 h of incubation [18].

Additionally, the higher activity of the bacteria in the GP medium and the increased consumption of glucose promoted higher production of gluconic acid (Tables 1 and 2, and Figures 1 and 2). Glucose dehydrogenase located in the cytoplasmic membrane of *K. xylinus* oxidizes glucose to gluconic acid, resulting in decreased conversion of glucose to BC [21]. In addition, gluconic acid promotes a significant decrease in the pH of the culture medium, which may hinder BC synthesis [21,34,35].

The times of greatest glucose decrease corresponded to the times of highest gluconic acid concentration in the medium, being 24.7 g/L in the GP medium and 15.4 g/L in the control medium. Note that, despite starting from similar values of glucose concentrations, the production of gluconic acid in the GP medium was 1.6 times higher than that in the control medium. A higher production of gluconic acid may be related to higher growth and activity of *K. xylinus* bacteria in the GP medium than those in the control medium. The presence of micronutrients and antioxidant components in bagasse are able to enhance bacterial growth by increasing glucose consumption and BC and gluconic acid production. The obtained results are in agreement with those observed in previous studies in which the presence of antioxidant components in the culture medium, such as vitamin C or lignosulfonate, improved BC production yields [33,36].

The synthesis of gluconic acid leads to a decrease in the pH of the culture medium that could hinder BC biosynthesis [21], as can be seen in the pH values of the media shown in Tables 1 and 2.

The pH of the media showed a sharp decrease on day 3 (medium control) and day 4 (medium GP) and a decrease of almost 50% of the initial glucose level that coincided with the maximum peaks of gluconic acid concentrations in both media. After 4 days of incubation time, the concentration of gluconic acid began to steadily decrease from 24.7 g/L to 9.7 g/L in the GP medium and from 15.4 g/L to 3.6 g/L in the control medium on day 10. In addition, the pH increased slightly from 3.5 to 3.8 in the GP medium and from 3.2 to 3.4 in the control medium as gluconic acid was consumed. The greatest increase in cellulose production occurred in the first few days (Tables 1 and 2), overlapping with the highest glucose consumption; thereafter, the increase was slower until reaching a maximum concentration of BC. This steady increase in BC concentration, complemented by a decrease in gluconic acid, indicated that the gluconic acid by-product could be used by *K. xylinus* to produce BC, as reported elsewhere [21,35].

The cellobiose content of both media increased slightly from 3.3 to 13.3 g/L and from 0.4 to 2.9 g/L as the incubation period advanced, being higher in the GP medium. Cellobiose is the structural unit of cellulose molecules and is also a product in the hydrolysis of cellulose [37]. This increase in cellobiose concentration could be related to the increase in the BC content in the plates, being higher in the samples with the GP medium. In the case of the monosaccharides, xylose and arabinose, they were only observed in the GP medium, as they were components derived from the hydrolysis of pectin and hemicelluloses. No variations in xylose concentration were observed throughout the incubation time. However, arabinose was consumed by day 6, when glucose concentrations reached minimum values (2.5 g/L) before being completely consumed. Thus, *K. xylinus* began to consume arabinose when glucose was close to being completely consumed.

The main drawback of acid hydrolysis is the generation of degradation products, such as furfural or 5-(hydroxymethyl)-2-furaldehyde (HMF). These by-products are microbial growth inhibitors and must be controlled under lethal concentrations to allow posterior fermentation. HMF is generated as a degradation product from glucose. Furfural is generated as a degradation product from pentose (xylose and arabinose) and can be found, for example, in grape seeds or in hemicellulosic portions [16,23,38]. The acid hydrolysis of the bagasse–potato mixture resulted in the appearance of HMF (241.6 mg/L) and furfural (7.9 mg/L). Note that, in both cases, as *K. xylinus* fermentation progressed, the concentration of both compounds decreased, indicating that the *K. xylinus* bacteria were able to consume HMF and furfural at the concentration level of the case study. HMF reached minimum values (0.7–1.3 mg/L) on day 6 of fermentation, and furfural was completely consumed by day 3. A similar behavior was observed in the control, in which only HMF (1.8 mg/L) was detected. Furfural was not detected due to the absence of hemicellulosic components in the medium. A significant inhibitory effect of HMF and furfural on the activity of *K. xylinus* was reported at a concentration of 2 g/L of HMF and 0.4 g/L of furfural [39]. In the GP medium, the concentration of both components was much lower, not having to affect BC production performance. Thus, no detoxification process of the hydrolysate was necessary before further use. However, no evidence of the *K. xylinus* consumption of HMF and furfural at low concentrations was found in the literature.

Regarding the drained BC weight, the pellicles obtained from the GP medium, the drained weight of the samples increased from 0.48 to 5.28 g as the incubation time increased, in accordance with the increase in the BC concentration. However, in the control medium, on day 2, the drained weight remained between 2.09 and 2.71 g during the whole fermentation time, in agreement with the observed data of dry weight and BC concentration, due to the interruption in BC production.

The swelling capacity data (Tables 1 and 2) indicate that, as the BC concentration increased, the moisture decreased. Thus, for dry pellicle values between 0.009 and 0.020 g, the swelling capacity remained between 13,352 and 28,169%. These values were reduced by more than eight times with increasing BC concentrations, obtaining a value of 3463% for 0.09 g dry samples. Likely, as the BC concentration increased, the sizes of the voids free to hold water decreased.

Overall, 6 days was enough to reach the maximum values of BC synthesis, because, after 6 days, the increase in BC is not relevant. At this period, in the GP medium, the estimated BC productivity was 0.028 g/L·h, the efficacy was 0.12 g/g and the yield was 0.13 g/g. However, in the control, the values were considerably lower, obtaining a productivity of 0.005 g/L, an efficacy of 0.03 g/g and a yield of 0.03 g/g. The significant difference in the parameters assessing BC synthesis showed enhanced production of BC in the medium from Garnacha bagasse and potato waste compared with the synthetic medium (control). The results indicate that the nutrients that were present in the proposed natural medium enhanced *K. xylinus* growth and BC production.

### 3.2. Characterization of Bacterial Cellulose Films

The physicochemical properties of BC pellicles obtained from the culture medium from the studied agro-food waste were evaluated. Unlike the films from the natural GP culture medium, the films synthesized in the control medium did not homogeneously form a film along the surface of the plate, as shown in Figure 3a. The films synthesized in the control culture medium were irregular and brittle, and it was not possible to dry them and obtain samples with homogeneous thickness for their characterization.

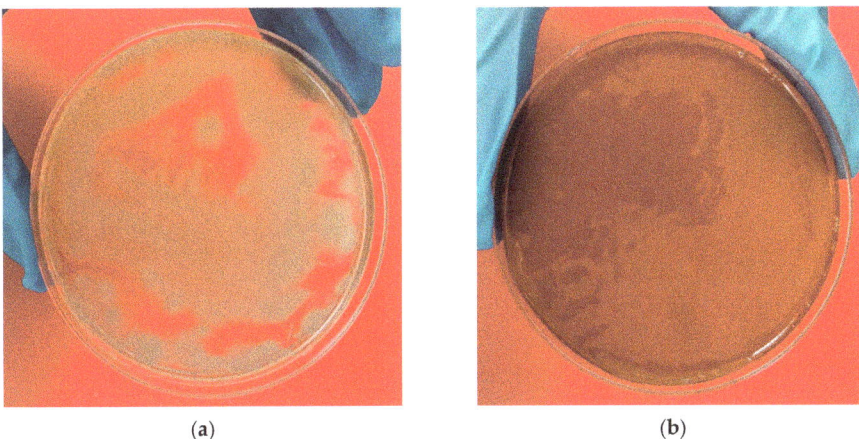

(a)  (b)

**Figure 3.** Images of bacterial cellulose films formed at the air–liquid interface of the culture medium: (**a**) Bacterial cellulose film growing in control culture medium showing non-homogeneous growth on the free surface of the Petri dish; (**b**) Bacterial cellulose film synthesized in bagasse–potato culture medium growing homogeneously covering the Petri dish surface.

Several Petri dishes with a diameter of 210 mm were prepared with GP culture medium and were incubated for 6 days at 30 °C under static conditions (Figure 3b). The BC membrane formed at the air–liquid interface was removed and was subsequently purified, washed and dried. Once the samples had been conditioned for at least 5 days at the specific relative humidity conditions for each test, they were measured.

The outstanding antioxidant capacity values of the films obtained confirmed the presence of the components of the GP culture medium in the polymer matrix despite the purification and washing processes to which the samples were subjected. The antioxidant capacity of film extracts obtained with ethanol and methanol was analyzed, observing TPC values of 0.31 and 1.32 mg GAE/g dried film, %DPPH$^{\bullet}$ of 57.24 and 78.00% and %ABTS$^{\bullet+}$ of 89.49 and 86.94%, respectively. The components with antioxidant capacity came mainly from bagasse hydrolysate with antioxidant capacity, as indicated above.

The obtained results for mechanical properties (tensile and puncture), water vapor permeability, moisture content at standard conditions of 57% RH and swelling capacity are presented in Table 3.

**Table 3.** Physicochemical properties of bacterial cellulose films synthesized in bagasse–potato culture media.

| Property | Value | Unit |
|---|---|---|
| Thickness | $0.010 \pm 0.001$ | mm |
| Tensile strength | $22.77 \pm 2.54$ | MPa |
| Percentage of elongation | $1.65 \pm 0.43$ | % |
| Young's Modulus | $910.46 \pm 48.66$ | MPa |
| Burst strength | $159.31 \pm 47.99$ | g |
| Distance to burst | $0.70 \pm 0.14$ | mm |
| Water vapor permeability | $3.40 \times 10^{-12} \pm 1.35 \times 10^{-13}$ | g/m·s·Pa |
| %W * | $10.07 \pm 1.55$ | % |
| Swelling capacity | $1053.27 \pm 45.59$ | % |

* Moisture content at equilibrium (57% RH).

Comparing the results shown in Table 3 with previously published results for BC films with a similar thickness (0.02 mm) obtained from *K. xylinus* from a commercial medium with 10% glucose and 1% yeast extract incubated for 8 days at 30 °C [26,28], slight differences were observed. The BC films obtained from the synthetic medium showed a TS value of 20.76 MPa, %E of 2.28% and a Young's modulus of 1043.88 MPa [28]. Considering the different incubation conditions, the films obtained from the GP medium showed higher breaking strength but lower deformation and elongation capacity. Likely, despite the purification process, some components of the bagasse–potato culture medium, such as phenolic compounds and antioxidants, remained embedded in the polymeric matrix of the BC, leading to a cross-linking effect. This effect could be seen most clearly in the puncture properties. Films from the GP medium showed higher BS and DB values (159.31 g and 0.70 mm) than those of films obtained from the synthetic medium (58.88 g and 0.39 mm) [26] despite having a lower thickness.

On the other hand, the obtained films showed lower permeability values of $3.40 \times 10^{-12}$ g/m·s·Pa compared with films previously obtained from the synthetic medium ($2.38 \times 10^{-11}$ g/m·s·Pa) [28]. The assembly pattern of BC fibers could be altered depending on the bagasse–potato components suspended in the medium, obtaining a modified BC pellicle at a morphological level. The compound in the medium could modify the conformation of the BC chains and lead to new physic interactions between the supplement and BC. The supplementation effect on the structural film properties could extend to the rest of the functional properties of the film, such as mechanical resistance and the permeability of the films [40]. In the present study, cellulose synthesis in the GP medium could result in a more closed and compact polymeric structure, which hindered the passage of water molecules through the matrix, resulting in lower WVP values. Moreover, these permeability values are lower than those shown by other polysaccharide matrices, extending the applicability of this BC matrix [41].

The moisture content of the samples obtained under standard conditions (10.07%) was higher than that previously observed in films obtained in synthetic media (1.82%) [28]. This difference could be mainly due to the presence of remains with hydrophilic properties in the culture medium. These hydrophilic remnants led to a higher swelling capacity of the BC matrix for the GP medium compared with the swelling capacity of a pure BC film (364.78%) with a thickness of 0.002 mm [26]. Additionally, the swelling capacity of the dried films (1053.27%) was lower than that observed in BC films removed directly from the Petri dishes (Tables 1 and 2). This is because, once BC films are dried, the structure closes and the physical interactions between the polymeric chains increase, not being able to reach swelling capacity values similar to those of a non-dried film [42].

The UV barrier properties of the BC samples from the GP culture medium were calculated from the transmittance values in the UV region. The transparency, opacity and color of the pellicles were determined from the transmittance values in the visible region. The samples showed a significant barrier capacity against UV radiation, giving mean values of percent transmittance ranging from 0.02 to 0.34%. The UV blocking capacity of the BC films were enhanced by structural modifications or components in the GP medium that remained between the polymer matrix. In previous work, BC films from synthetic media showed UV barrier values ranging from 1.35 to 5.61% [4]. These remains of the GP medium trapped between the cellulose chains also modified the opacity and color properties of the films. The transparency of the film (26.97 ± 51.47) remained at values similar to those previously observed in synthetic BC medium films [4], but the opacity (214.57 ± 61.95) increased more than 10-fold. In addition, the lightness of the samples (L* value of 14.16 ± 11.91) decreased, and the redness (a* value of 4.84 ± 1.63) and yellowness (b* value of 11.05 ± 6.21) increased.

## 4. Conclusions

This work demonstrates that it is possible to improve the production and yield of BC from a natural medium based on bagasse and non-commercial potato waste. The proposed low-cost culture medium could be an alternative to give added value to the by-products of the agri-food industry and to obtain a raw material with potential use in the food, cosmetic, medical and pharmaceutical industries. Nutrients of the culture medium, due to the presence of bagasse, increased BC production yield compared with the control synthetic medium. The enriched medium also promoted increased activity of the bacteria, converting glucose to gluconic acid. The production of gluconic acid reduced the availability of glucose and lowered the pH of the medium, hindering BC synthesis. The ability of *K. xylinus* to consume HMF and F at low concentrations was observed, which raises interesting applications that deserve to be deeply studied in future work.

The obtained films showed permeability values in orders of one and two magnitudes lower than the permeability properties of most of the other polysaccharide-based films. In addition, components with antioxidant capacity from Garnacha bagasse were retained between the BC chains during polymer synthesis, providing antioxidant capacity and functionality to the films.

The UV radiation blocking capacity of the samples, together with their antioxidant capacity, edibility and water insolubility, make BC films from GP media a material with potential applications in direct food contact to retard oxidation reactions that may modify the organoleptic properties of the product.

**Author Contributions:** Conceptualization, P.C.; methodology, P.C. and G.P.; investigation, P.C.; writing—original draft preparation, P.C.; writing—review and editing, G.P. and M.V.; supervision, M.V.; funding acquisition, M.V. All authors have read and agreed to the published version of the manuscript.

**Funding:** The authors appreciate the funding support of Xunta de Galicia, within the postdoctoral fellowship granted to Patricia Cazón Díaz (No. ED481B-2021–040).

**Institutional Review Board Statement:** Not applicable.

**Data Availability Statement:** The data presented in this study are available on request from the corresponding author.

**Acknowledgments:** The use of RIAIDT-USC analytical facilities is acknowledged. In addition, the authors would like to thank the researcher María Belén García Gómez for her assistance with the wine company and for providing the Garnacha bagasse samples.

**Conflicts of Interest:** The authors declare that they have no known competing financial interest or personal relationships that could have appeared to influence the work reported in this paper.

## References

1.    Otoni, C.G.; Azeredo, H.M.C.; Mattos, B.D.; Beaumont, M.; Correa, D.S.; Rojas, O.J. The Food–Materials Nexus: Next Generation Bioplastics and Advanced Materials from Agri-Food Residues. *Adv. Mater.* **2021**, *33*, 2170342. [CrossRef]

2.  Kovalcik, A.; Pernicova, I.; Obruca, S.; Szotkowski, M.; Enev, V.; Kalina, M.; Marova, I. Grape Winery Waste as a Promising Feedstock for the Production of Polyhydroxyalkanoates and Other Value-Added Products. *Food Bioprod. Process.* **2020**, *124*, 1–10. [CrossRef]

3.  Hussain, Z.; Sajjad, W.; Khan, T.; Wahid, F. Production of Bacterial Cellulose from Industrial Wastes: A Review. *Cellulose* **2019**, *1*, 2895–2911. [CrossRef]

4.  Cazon, P.; Velázquez, G.; Vázquez, M. Bacterial Cellulose Films: Evaluation of the Water Interaction. *Food Packag. Shelf Life* **2020**, *25*, 100526. [CrossRef]

5.  Picheth, G.F.; Pirich, C.L.; Sierakowski, M.R.; Woehl, M.A.; Sakakibara, C.N.; de Souza, C.F.; Martin, A.A.; da Silva, R.; de Freitas, R.A. Bacterial Cellulose in Biomedical Applications: A Review. *Int. J. Biol. Macromol.* **2017**, *104*, 97–106. [CrossRef]

6.  Ullah, H.; Santos, H.A.; Khan, T. Applications of Bacterial Cellulose in Food, Cosmetics and Drug Delivery. *Cellulose* **2016**, *23*, 2291–2314. [CrossRef]

7.  Cazón, P.; Vázquez, M. Bacterial Cellulose as a Biodegradable Food Packaging Material: A Review. *Food Hydrocoll.* **2021**, *113*, 106530. [CrossRef]

8.  Torres, F.G.; Troncoso, O.P.; Gonzales, K.N.; Sari, R.M.; Gea, S. Bacterial Cellulose-based Biosensors. *Med. Devices Sens.* **2020**, *3*, e10102. [CrossRef]

9.  Portilla Rivera, O.M.; Saavedra Leos, M.D.; Solis, V.E.; Domínguez, J.M. Recent Trends on the Valorization of Winemaking Industry Wastes. *Curr. Opin. Green Sustain. Chem.* **2021**, *27*, 100415. [CrossRef]

10. Cerrutti, P.; Roldán, P.; García, R.M.; Galvagno, M.A.; Vázquez, A.; Foresti, M.L. Production of Bacterial Nanocellulose from Wine Industry Residues: Importance of Fermentation Time on Pellicle Characteristics. *J. Appl. Polym. Sci.* **2016**, *133*, 43109. [CrossRef]

11. Carreira, P.; Mendes, J.A.S.; Trovatti, E.; Serafim, L.S.; Freire, C.S.R.; Silvestre, A.J.D.; Neto, C.P. Utilization of Residues from Agro-Forest Industries in the Production of High Value Bacterial Cellulose. *Bioresour. Technol.* **2011**, *102*, 7354–7360. [CrossRef]

12. Ogrizek, L.; Lamovšek, J.; Čuš, F.; Leskovšek, M.; Gorjanc, M. Properties of Bacterial Cellulose Produced Using White and Red Grape Bagasse as a Nutrient Source. *Processes* **2021**, *9*, 1088. [CrossRef]

13. Rani, M.U.; Udayasankar, K.; Appaiah, K.A.A. Properties of Bacterial Cellulose Produced in Grape Medium by Native Isolate *Gluconacetobacter* sp. *J. Appl. Polym. Sci.* **2011**, *120*, 2835–2841. [CrossRef]

14. Wu, D. Recycle Technology for Potato Peel Waste Processing: A Review. *Procedia Environ. Sci.* **2016**, *31*, 103–107. [CrossRef]

15. Maroušek, J.; Rowland, Z.; Valášková, K.; Král, P. Techno-Economic Assessment of Potato Waste Management in Developing Economies. *Clean Technol. Environ. Policy* **2020**, *22*, 937–944. [CrossRef]

16. Guerra-Rodríguez, E.; Portilla-Rivera, O.M.; Ramírez, J.A.; Vázquez, M. Modelling of the Acid Hydrolysis of Potato (*Solanum Tuberosum*) for Fermentative Purposes. *Biomass Bioenergy* **2012**, *42*, 59–68. [CrossRef]

17. Kot, A.M.; Pobiega, K.; Piwowarek, K.; Kieliszek, M.; Błażejak, S.; Gniewosz, M.; Lipińska, E. Biotechnological Methods of Management and Utilization of Potato Industry Waste—A Review. *Potato Res.* **2020**, *63*, 431–447. [CrossRef]

18. Abdelraof, M.; Hasanin, M.S.; El-Saied, H. Ecofriendly Green Conversion of Potato Peel Wastes to High Productivity Bacterial Cellulose. *Carbohydr. Polym.* **2019**, *211*, 75–83. [CrossRef]

19. Vazquez, M.; Martin, A.M. Mathematical Model for Phaþia Rhodozyma Growth Using Peat Hydrolysates as Substrate. *J. Sci. Food Agric.* **1998**, *76*, 481–487. [CrossRef]

20. Salari, M.; Sowti Khiabani, M.; Rezaei Mokarram, R.; Ghanbarzadeh, B.; Samadi Kafil, H. Preparation and Characterization of Cellulose Nanocrystals from Bacterial Cellulose Produced in Sugar Beet Molasses and Cheese Whey Media. *Int. J. Biol. Macromol.* **2019**, *122*, 280–288. [CrossRef]

21. Kuo, C.-H.; Teng, H.-Y.; Lee, C.-K. Knock-out of Glucose Dehydrogenase Gene in Gluconacetobacter Xylinus for Bacterial Cellulose Production Enhancement. *Biotechnol. Bioprocess. Eng.* **2015**, *20*, 18–25. [CrossRef]

22. Singh, O.V.; Singh, R.P. Bioconversion of Grape Must into Modulated Gluconic Acid Production by Aspergillus Niger ORS-4·410. *J. Appl. Microbiol.* **2006**, *100*, 1114–1122. [CrossRef] [PubMed]

23. Guerra-Rodríguez, E.; Portilla-Rivera, O.M.; Jarquín-Enríquez, L.; Ramírez, J.A.; Vázquez, M. Acid Hydrolysis of Wheat Straw: A Kinetic Study. *Biomass Bioenergy* **2012**, *36*, 346–355. [CrossRef]

24. Flórez, M.; Cazón, P.; Vázquez, M. Active Packaging Film of Chitosan and Santalum Album Essential Oil: Characterization and Application as Butter Sachet to Retard Lipid Oxidation. *Food Packag. Shelf Life* **2022**, *34*, 100938. [CrossRef]

25. Flórez, M.; Cazón, P.; Vázquez, M. Antioxidant Extracts of Nettle (Urtica Dioica) Leaves: Evaluation of Extraction Techniques and Solvents. *Molecules* **2022**, *27*, 6015. [CrossRef]

26. Cazón, P.; Vázquez, M.; Velazquez, G. Composite Films with UV-Barrier Properties Based on Bacterial Cellulose Combined with Chitosan and Poly(Vinyl Alcohol): Study of Puncture and Water Interaction Properties. *Biomacromolecules* **2019**, *20*, 2084–2095. [CrossRef]

27. Cazon, P.; Vazquez, M.; Velazquez, G. Composite Films of Regenerate Cellulose with Chitosan and Polyvinyl Alcohol: Evaluation of Water Adsorption, Mechanical and Optical Properties. *Int. J. Biol. Macromol.* **2018**, *117*, 235–246. [CrossRef]

28. Cazón, P.; Velázquez, G.; Vázquez, M. Characterization of Bacterial Cellulose Films Combined with Chitosan and Polyvinyl Alcohol: Evaluation of Mechanical and Barrier Properties. *Carbohydr. Polym.* **2019**, *216*, 72–85. [CrossRef]

29. Cazon, P.; Velázquez, G.; Vázquez, M. UV-Protecting Films Based on Bacterial Cellulose, Glycerol and Polyvinyl Alcohol: Effect of Water Activity on Barrier, Mechanical and Optical Properties. *Cellulose* **2020**, *27*, 8199–8213. [CrossRef]

30. González-Centeno, M.R.; Rosselló, C.; Simal, S.; Garau, M.C.; López, F.; Femenia, A. Physico-Chemical Properties of Cell Wall Materials Obtained from Ten Grape Varieties and Their Byproducts: Grape Pomaces and Stems. *LWT Food Sci. Technol.* **2010**, *43*, 1580–1586. [CrossRef]
31. Spigno, G.; Tramelli, L.; De Faveri, D.M. Effects of Extraction Time, Temperature and Solvent on Concentration and Antioxidant Activity of Grape Marc Phenolics. *J. Food Eng.* **2007**, *81*, 200–208. [CrossRef]
32. Martelo-Vidal, M.J.; Vázquez, M. Determination of Polyphenolic Compounds of Red Wines by UV-VIS-NIR Spectroscopy and Chemometrics Tools. *Food Chem.* **2014**, *158*, 28–34. [CrossRef]
33. Keshk, S. Vitamin C Enhances Bacterial Cellulose Production in Gluconacetobacter Xylinus. *Carbohydr. Polym.* **2014**, *99*, 98–100. [CrossRef]
34. Lin, S.P.; Loira Calvar, I.; Catchmark, J.M.; Liu, J.R.; Demirci, A.; Cheng, K.C. Biosynthesis, Production and Applications of Bacterial Cellulose. *Cellulose* **2013**, *20*, 2191–2219. [CrossRef]
35. Reiniati, I.; Hrymak, A.N.; Margaritis, A. Recent Developments in the Production and Applications of Bacterial Cellulose Fibers and Nanocrystals. *Crit. Rev. Biotechnol.* **2017**, *37*, 510–524. [CrossRef]
36. Keshk, S.; Sameshima, K. Influence of Lignosulfonate on Crystal Structure and Productivity of Bacterial Cellulose in a Static Culture. *Enzym. Microb. Technol.* **2006**, *40*, 4–8. [CrossRef]
37. Zhao, Y.; Wu, B.; Yan, B.; Gao, P. Mechanism of Cellobiose Inhibition in Cellulose Hydrolysis by Cellobiohydrolase. *Sci. China Ser. C Life Sci.* **2004**, *47*, 18–24. [CrossRef]
38. Yedro, F.M.; García-Serna, J.; Cantero, D.A.; Sobrón, F.; Cocero, M.J. Hydrothermal Fractionation of Grape Seeds in Subcritical Water to Produce Oil Extract, Sugars and Lignin. *Catal. Today* **2015**, *257*, 160–168. [CrossRef]
39. Kim, H.; Son, J.; Lee, J.; Yoo, H.Y.; Lee, T.; Jang, M.; Oh, J.M.; Park, C. Improved Production of Bacterial Cellulose through Investigation of Effects of Inhibitory Compounds from Lignocellulosic Hydrolysates. *GCB Bioenergy* **2021**, *13*, 436–444. [CrossRef]
40. Dayal, M.S.; Catchmark, J.M. Mechanical and Structural Property Analysis of Bacterial Cellulose Composites. *Carbohydr. Polym.* **2016**, *144*, 447–453. [CrossRef]
41. Cazon, P.; Velazquez, G.; Ramírez, J.A.; Vázquez, M. Polysaccharide-Based Films and Coatings for Food Packaging: A Review. *Food Hydrocoll.* **2017**, *68*, 136–148. [CrossRef]
42. Lin, S.-B.; Hsu, C.-P.; Chen, L.-C.; Chen, H.-H. Adding Enzymatically Modified Gelatin to Enhance the Rehydration Abilities and Mechanical Properties of Bacterial Cellulose. *Food Hydrocoll.* **2009**, *23*, 2195–2203. [CrossRef]

Article

# Bioinspired Electropun Fibrous Materials Based on Poly-3-Hydroxybutyrate and Hemin: Preparation, Physicochemical Properties, and Weathering

Polina M. Tyubaeva [1], Ivetta A. Varyan [1,2], Anna K. Zykova [1], Alena Yu. Yarysheva [3], Pavel V. Ivchenko [3,*], Anatoly A. Olkhov [1,2] and Olga V. Arzhakova [3]

1    Academic Department of Technology and Chemistry of Innovative Materials, Plekhanov University of Economics, Stremyanny Per. 36, Moscow 117997, Russia
2    Department of Biological and Chemical Physics of Polymers, Emanuel Institute of Biochemical Physics, Russian Academy of Sciences, ul. Kosygina 4, Moscow 119334, Russia
3    Faculty of Chemistry, Lomonosov Moscow State University, Leninskie Gory 1/3, Moscow 119991, Russia
*    Correspondence: phpasha1@yandex.ru

Citation: Tyubaeva, P.M.; Varyan, I.A.; Zykova, A.K.; Yarysheva, A.Y.; Ivchenko, P.V.; Olkhov, A.A.; Arzhakova, O.V. Bioinspired Electropun Fibrous Materials Based on Poly-3-Hydroxybutyrate and Hemin: Preparation, Physicochemical Properties, and Weathering. *Polymers* 2022, *14*, 4878. https://doi.org/10.3390/polym14224878

Academic Editor: Piotr Bulak

Received: 23 September 2022
Accepted: 9 November 2022
Published: 12 November 2022

Publisher's Note: MDPI stays neutral with regard to jurisdictional claims in published maps and institutional affiliations.

**Abstract:** The development of innovative fibrous materials with valuable multifunctional properties based on biodegradable polymers and modifying additives presents a challenging direction for modern materials science and environmental safety. In this work, high-performance composite fibrous materials based on semicrystalline biodegradable poly-3-hydroxybutyrate (PHB) and natural iron-containing porphyrin, hemin (*Hmi*) were prepared by electrospinning. The addition of *Hmi* to the feed PHB mixture (at concentrations above 3 wt.%) is shown to facilitate the electrospinning process and improve the quality of the electrospun PHB/*Hmi* materials: the fibers become uniform, their average diameter decreases down to 1.77 μm, and porosity increases to 94%. Structural morphology, phase composition, and physicochemical properties of the *Hmi*/PHB fibrous materials were studied by diverse physicochemical methods, including electronic paramagnetic resonance, optical microscopy, scanning electron microscopy, and energy-dispersive X-ray spectroscopy, elemental analysis, differential scanning calorimetry, Fourier-transformed infrared spectroscopy, mechanical analysis, etc. The proposed nonwoven *Hmi*/PHB composites with high porosity, good mechanical properties, and retarded biodegradation due to high antibacterial potential can be used as high-performance and robust materials for biomedical applications, including breathable materials for wound disinfection and accelerated healing, scaffolds for regenerative medicine and tissue engineering.

**Keywords:** biodegradable polymers; poly(3-hydroxybutyrate); hemin; electrospinning; outdoor weathering; new technologies

## 1. Introduction

At the present time, the mainstream direction of modern materials science and technology is oriented toward the development of innovative and ecologically friendly materials based on natural components and biodegradable polymers with task-oriented multifunctional properties presents. In this connection, the family of biodegradable polyhydroxyalkanoates, including poly(hydroxybutyrate) (PHB) is of special interest due to their potential as biomedical materials for therapeutic applications. PHB seems to be a promising candidate for the development of composite materials for biomedical applications as this biopolymer is characterized by high melting temperature, high degree of crystallinity, and low permeability to oxygen, water, and carbon dioxide [1]. This biodegradable polymer can be synthesized from renewable natural sources and is known to be biocompatible with the human body [2].

The introduction of various additives to PHB allows the preparation of the composite PHB-based materials with desired functional properties (nanoparticles [3], carbon nanotubes [4], catalysts, and enzymes [5]). Nowadays, special interest is focused on the molecular complexes of porphyrins and related materials for their applied use as superconductors [6–8], in analytical chemistry [9,10], photonics [11,12], gene therapy [13,14], biomedicine [15–17]. Progress in the synthesis and chemistry of tetrapyrroles has triggered the ever-growing attention to tetrapyrrole complexes as an effective modifying additive for polymeric materials, including numerous analogs of natural systems such as porphyrins, texaphyrins, chlorins, corroles, and others, applied for biomedical purposes [18–22].

Hemin is the iron-containing porphyrin, which is endogenously produced in the human body, for example, during the turnover of old red blood cells. The unique properties and benefits of hemin have been widely discussed in the literature [23–32]. Evidently, hemin is biocompatible with living organisms [23] and can be used for the stabilization of protein molecules on polymer carriers, and hemin-containing materials can be successfully used for diverse biomedical applications [24], including, for example, the development of safe containers for drug delivery systems [25]. Moreover, peptides can be attached to the surface of hemin-containing materials [26,27]. Hemin shows a well-pronounced catalytic activity [28] and can be effectively used in anticoagulants [29]. Hemin is characterized by high thermal stability, thus lifting the challenging problems related to the processing of polymers and the preparation of sustainable polymer nanocomposite materials [30]. In addition, hemin also shows high antimicrobial activity against Gram-positive *S. aureus* [31], which is a significant advantage in the creation of biomedical materials. Finally, the benefits of hemin are concerned with its natural origin biocompatibility with living organisms, and simple and facile procedure of synthesis [32]. Hence, hemin was selected as an efficient additive for the modification of the PHB materials and preparation of new *Hmi*/PHB materials with task-oriented functional properties.

Electrospinning (ES) is credited as a reliable and efficient method for the preparation of high-performance nonwoven materials based on a wide range of polymers [33]. This method provides the formation of an interconnected network of ultrathin fibers by drawing a jet of the polymer solution under the action of physical forces. Electrospinning offers powerful tools for the preparation of diverse composite materials when a functional additive is dissolved in the polymer feed solution [34–37]: poly(3-hydroxybutyrate-co-3-hydroxyvalerate)/polylactide [38], poly(butylene adipate succinate)/polylactide [39], poly(butylene adipate terephthalate)/polylactide [40], poly($\varepsilon$-caprolactone)/poly (3-hydroxybutyrate) [41]. This ES approach allows one to control the content of additives in the composite, phase composition, and distribution of the additive within the fibrous materials.

This work addresses the development of a facile and sustainable method for the preparation of high-performance innovative biocompatible *Hmi*/PHB materials by electrospinning, characterization of their morphology, phase composition, and mechanical properties by diverse physicochemical methods (electronic paramagnetic resonance (EPR), optical microscopy, scanning electron microscopy (SEM) and energy-dispersive X-ray spectroscopy (EDX) elemental analysis, differential scanning calorimetry (DSC), Fourier-transformed infrared (FTIR) spectroscopy, mechanical analysis, etc.), investigation of their stability upon outdoor weathering tests, and description of the potential applied benefits of the resultant *Hmi*/PHB materials for diverse applications.

## 2. Materials and Methods

### 2.1. Materials

In this work, semicrystalline biodegradable polymer, poly(3-hydroxybutyrate) (PHB) (trade mark 16F series, BIOMER, Frankfurt, Germany) was used; molecular weight—206 kDa, density—1.5 g/cm$^3$ and degree of crystallinity—59% (Figure 1A). Hemin (a tetrapyrrole complex from the class of natural porphyrins) was used as a modifying additive (Figure 1B). Hemin with the coordination complex of iron (oxidation state III) was obtained from the bovine (Aldrich Sigma, Saint Louis, MO, USA).

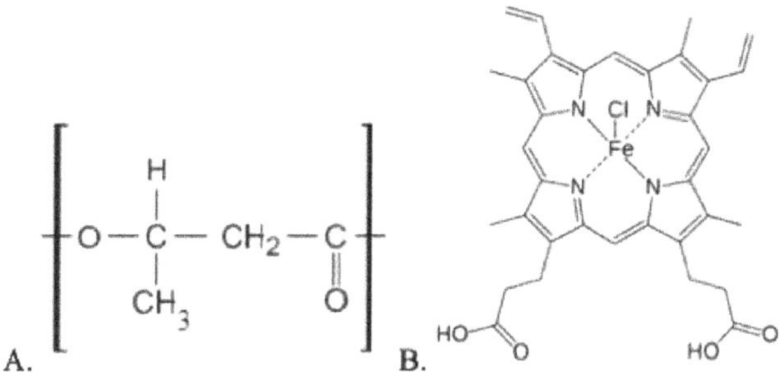

**Figure 1.** Structural formulae of PHB (**A**) and hemin (**B**).

*2.2. Preparation of Electrospun Materials*

In this work, polymer nanofibrous materials were prepared by electrospinning (ES) technique using a single-capillary laboratory unit with a capillary diameter of 0.1 mm (EFV-1, Moscow, Russia) (Supplementary Files, Figure S1). As a result of electrospinning, polymer fibers are cured due to the complete evaporation of the solvent. For the preparation of feed solutions, finely dispersed PHB powder was dissolved in chloroform at a temperature of 60 °C. Hemin was dissolved in N,N-dimethylformamide at 25 °C and homogenized with the PHB solution. The content of PHB in the solution was 7 wt.%, and the content of hemin was 1, 3, and 5 wt.% of the PHB. The conditions of the ES process were the following: voltage 17–20 kV, the distance between the electrodes—190–200 mm, the gas pressure on the solution—10–15 kg(f) cm$^{-2}$. The electrical conductivity of the feed solution was 10–14 μS cm$^{-1}$ and the viscosity of the hemin/PHB solution was 1.4–1.9 Pa s (viscosity of the 7 wt.% PHB solution in chloroform was 1 Pa s) and the flow rate of the solution was 0.15–0.20 cm$^3$ min$^{-1}$ (flow rate of the 7 wt.% PHB solution in chloroform was 0.45 cm$^3$ min$^{-1}$).

*2.3. Methods*

2.3.1. Optical Microscopy

Primary data on topography and the general appearance of the samples were collected using an Olympus BX43 optical microscope (Tokyo, Japan) in transmission mode.

2.3.2. Scanning Electron Microscopy (SEM)

The morphology of the samples was examined by SEM observations using a Tescan VEGA3 microscope (Wurttemberg, Czech Republic). Prior to SEM observations, the surface of the test samples was decorated with a thin platinum layer.

2.3.3. Estimation of Structural Parameters

The morphology of the fibrous materials was studied using the Olympus Stream Basic software (Tokyo, Japan). The average diameter of individual fibers was estimated for each fiber on the z-stack at five different points. The density δ of the samples reads as:

$$\delta = \frac{m}{l \times B \times b} \tag{1}$$

where $m$ is the weight of the sample; $l$ is the length; $B$ is the width; $b$ is the thickness.

The porosity of the fibrous electrospun materials was estimated as the ratio between the volume of pores and the overall volume of the sample:

$$W = (1 - \frac{V_f}{V_s}) \times 100, \% \tag{2}$$

where $W$ is the porosity of the nonwoven sample, $V_f$ and $V_s$ stand for the volume of polymer as fibers in the nonwoven sample and the nonwoven sample, respectively.

2.3.4. Energy-Dispersive X-ray Spectroscopy (EDX)

The elemental composition of the samples (oxygen, chlorine, and iron) was studied by the method of energy-dispersive X-ray spectroscopy (EDX) using a Tescan VEGA3 instrument (Wurttemberg, Czech Republic).

2.3.5. FTIR Spectroscopy

The chemical composition of initial, modified, and weathered materials were studied using an FTIR Lumos BRUKER (Berlin, Germany) spectrometer at a temperature of $(22 \pm 2)$ °C in the range of wavenumbers of $4000 \leq \nu \leq 600$ cm$^{-6}$ in the mode of frustrated total internal reflection (FTIR) on the diamond crystal [42]. The intensity of several absorption peaks related to polymer degradation (functional groups) was estimated [43,44].

2.3.6. Differential Scanning Calorimetry (DSC)

The thermal properties of the test samples were studied using a Netzsch 214 Polyma thermal analyzer (Selb, Germany); the heating rate was 10 °C/min. The samples were heated from 20 °C to 220 °C and then cooled down to 20 °C. The heat of fusion was calculated using the NETZSCH Proteus software (Selb, Germany). The weight of the samples was ~7 mg. The data were analyzed according to the standard procedure [45]. The degree of crystallinity of the samples $\chi$ reads as:

$$\chi = \Delta H_m / \Delta H_{m,100\%} \times 100\% \tag{3}$$

where $\Delta H_m$ is the heat of fusion; $\Delta H_{m,100\%}$ is the heat of fusion of the ideal crystal of PHB (146 J/g [46]).

2.3.7. Electronic Paramagnetic Resonance (EPR)

The EPR spectra were collected using an EPR-V automated spectrometer (Moscow, Russia). Microwave power in the resonator cavity was below 7 mW. Stable nitroxyl radical (2,2,6,6-tetramethyl piperidine-1-oxyl (TEMPO) radical moiety) was used as a probe. The TEMPO radical was introduced into the polymer samples from the gas phase at 60 °C. The content of the TEMPO radicals in the polymer samples was estimated using Bruker WinEPR and SimFonia software (Berlin, Germany) (CCl$_4$ with the radical concentration below $10^{-3}$ mol L$^{-1}$ as a reference). Correlation times of the probe rotation were estimated from the corresponding EPR spectra.

The EPR spectra of the TEMPO spin probe in the slow-motion region ($\tau > 10^{-10}$ s) were analyzed in the terms of the model of isotropic Brownian rotation using the NLSL software [47]. Correlation times of the probe rotation $\tau$ in the fast-motion region ($5 \times 10^{-11} < \tau < 10^{-9}$ s) were estimated from the corresponding EPR spectra as:

$$\tau = 6.65 \times 10^{-10}(\sqrt{I_+/I_-} - 1) \times \Delta H_+ \tag{4}$$

where $\Delta H_+$ is the half-width of the spectrum component in the weak field, and $I_+/I_-$ is the ratio of the intensities of the components in weak and strong fields, respectively [48].

2.3.8. Mechanical Tests

Mechanical properties (tensile strength, elongation at break) of the test samples were estimated using a Devotrans DVT GP UG 5 tensile compression testing machine (Istanbul, Turkey) according to ASTM D5035-11.

2.3.9. Weathering Experiments upon Outdoor Exposure

The weathering experiments were performed according to the ASTM G7/G7M-21 ("Standard Practice for Natural Weathering of Materials"). The samples were subjected to outdoor exposure according to standard protocol D1435-13 ("Standard Practice for Outdoor Weathering of Plastics"). According to the protocol, the wooden weathering racks were tilted at an angle of 90° with respect to the horizontal line.

The dimensions of the samples were 40 mm × 40 mm × 0.3 mm (length, width, and thickness, respectively). Similar samples were placed as blank samples in a dark and dry place. The specimens were fastened with white cotton threads; one side of the specimens was fastened with adhesive tape. The duration of each weathering experiment was 3 months (June–August; temperature average 21 °C; humidity average 75.7%; geographical location: 55° 45′ north latitude; 37° 37′ east longitude).

### 2.3.10. Statistical Analysis

All measurements were performed, at least, three times for each sample. Mechanical tests, weathering experiments, and measurements of structural characteristics were performed for 8–10 samples. A mean value and standard deviation were estimated using Data Analysis by the Origin Pro program (OriginLab Corporation, Northampton, MA, USA).

## 3. Results

Electrospinning is a well-known and reliable procedure for the development of a fibrous structure [49,50]. This method makes it possible to control the supramolecular structure and morphology of individual fibers as well as the morphology of the system as a whole [51]. The introduction of modifying additives even at low concentrations (below 5 wt.%) provides a marked effect on the formation of the structure of the resultant materials at all structural levels.

### 3.1. Morphology Structure and Properties of Hemin/PHB Fibrous Materials

Nonwoven materials prepared by electrospinning are characterized by well-developed morphology which is described by several key parameters, including density, the average diameter of individual fibers, porosity, and pore dimensions. Figure 2 shows the optical micrographs illustrating the visual appearance of the *Hmi*/PHB samples with different content of *Hmi*.

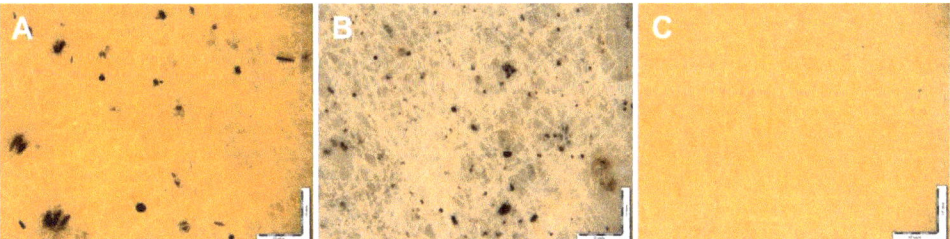

**Figure 2.** The micrographs of *Hmi*/PHB fibrous nanocomposites with different content of hemin: 1 wt.% (**A**), 3 wt.% (**B**), and 5 wt.% (**C**).

As follows from Figure 2, the *Hmi*/PHB fibrous samples contain big-sized dark inclusions which are accommodated on the surface of individual fibers of the nonwoven sample (Figure 2A,B). The dimensions of the inclusions are equal to 4–32 μm and 0.7–17 μm for *Hmi*/PHB containing 1 and 3 wt.% of *Hmi*, respectively. Noteworthy is that the *Hmi*/PHB samples containing 5 wt.% *Hmi* are seen to be uniform and free of big-sized inclusions (Figure 2C). This fact indirectly suggests that electrospinning from the feed solutions containing 3–5 wt.% of *Hmi* is enhanced and provides the preparation of more uniform nonwoven *Hmi*/PHB materials. Analysis of the micrographs in Figure 2 allows estimation of the average diameter of individual fibers of the *Hmi*/PHB nonwoven materials. With increasing the content of *Hmi* to 5 wt.%, the average diameter of the fibers decreases from 3.5 μm for the neat PHB down to 1.77 μm for the *Hmi*/PHB containing 5 wt.% of *Hmi*.

Noteworthy is that, as the content of *Hmi* in the *Hmi*/PHB composites increases, porosity of nonwoven materials increases from 80% (for neat PHB) to 94%. Hence, this evidence allows one to conclude that electrospinning of the feed solutions with the maximum content of *Hmi* leads to the preparation of nonwoven *Hmi*/PHB materials with uniform structure and high porosity. Structural characteristics of the nonwoven *Hmi*/PHB materials are listed in Table 1. The histogram illustrating the size distribution of PHB and *Hmi*/PHB nonwoven composites is given in the Supplementary Files, Figure S2.

**Table 1.** Structural characteristics of PHB and *Hmi*/PHB nonwoven composites.

| Content of *Hmi* % | Density (Mean $\pm$ SD, $n = 10$) g/cm$^3$ | Average Diameter of Fibers (Mean $\pm$ SD, $n = 100$) $\mu$m | Porosity (Mean $\pm$ SD, $n = 10$) % |
|---|---|---|---|
| 0 | $0.30 \pm 0.01$ | $3.50 \pm 0.08$ | $80 \pm 2.0$ |
| 1 | $0.20 \pm 0.02$ | $2.06 \pm 0.07$ | $92 \pm 1.5$ |
| 3 | $0.20 \pm 0.01$ | $1.77 \pm 0.04$ | $92 \pm 1.5$ |
| 5 | $0.17 \pm 0.01$ | $1.77 \pm 0.04$ | $94 \pm 1.2$ |

Hence, the presence of 5 wt.% *Hmi* in the *Hmi*/PHB feed mixture is associated with a lower density of the feed solution and facilitates the ES process due to better-organized movement through the jet. As a result, pore size decreases, and porosity increases. This is due to the fact that the diameter of the PHB fibers is higher than that of the Hemin-modified fibers. Thicker PHB fibers are also characterized by the presence of large defects and their diameter exceeds the average diameter of the fibers by 3–10 times. As a result, upon ES, thicker fibers are located at a higher distance from each other, forming large pores, and this effect is the consequence of low electrical conductivity of the spinning solution. However, as Hemin is added into the feed solution, the number of big-sized defects on the surface of fibers decreases, the diameter of fibers is reduced, thinner fibers are packed in a more uniform mode, and the distance between the neighboring fibers is smaller, thus leading to smaller pores. In other words, when the density of the material is reduced (Table 1), the resultant material is characterized by a higher content of air layer between well-cured fibers without bonds, smudges, and thicker regions.

The fine structure of the fibrous *Hmi*/PHB materials was studied in more detail by SEM observations. Figure 3 shows the corresponding SEM images. As follows from Figure 3, the fibrous materials based on neat PHB and *Hmi*/PHB with a low content of *Hmi* (below 3 wt.%) are seen to be non-uniform; the fibers contain joints, crimps, and big-sized elements as spindles. The average size of structural imperfections ranges between 20 and 30 $\mu$m along the longitudinal direction and between 15–25 $\mu$m in the perpendicular direction. With increasing the content of *Hmi* in the *Hmi*/PHB system, the diameter of fibers markedly decreases, and the fibers within nonwoven materials are seen to be more uniform. Noteworthy is that, for the *Hmi*/PHB systems containing 5 wt.% of *Hmi*, the individual fibers are nearly free of structural defects and imperfections. Both roughness on the fiber surface almost completely, and spindle-shaped thickenings disappeared at 5 wt.% content of Hemin.

To gain a deeper insight into the fine structure of the *Hmi*/PHB materials, elemental analysis, and distribution of *Hmi* additive within the fibers, the *Hmi*/PHB materials were studied by the EDX method (Figure 4).

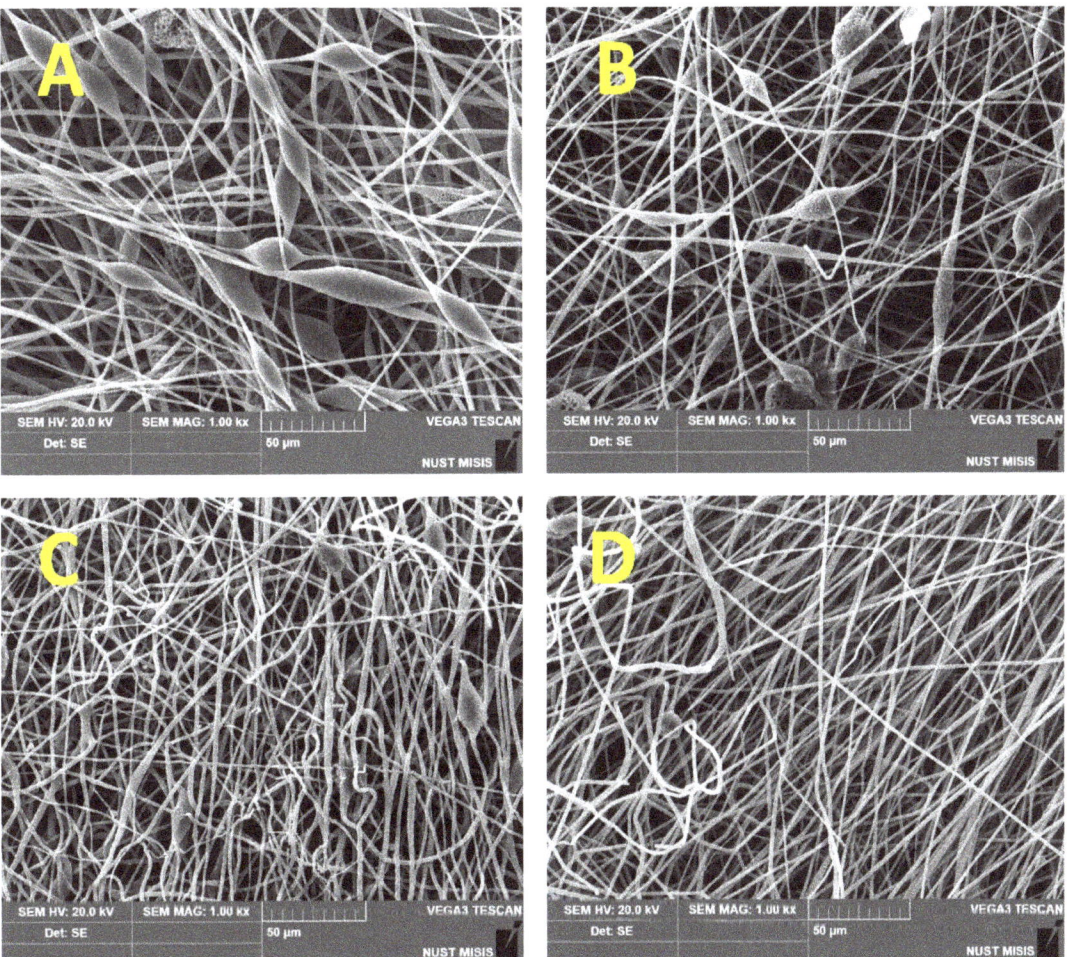

**Figure 3.** SEM micrographs of PHB and *Hmi*/PHB nonwoven materials with different content of *Hmi*: (**A**) neat PHB, (**B**) *Hmi*/PHB with 1 wt.% of *Hmi*, (**C**) *Hmi*/PHB with 3 wt.%, (**D**) *Hmi*/PHB with 5 wt.% of *Hmi*.

According to the EDX analysis, hemin is incorporated into fibers and uniformly distributed within the polymer structure on the surface and in the bulk. Iron and chlorine atoms serve as probes for the characterization of the distribution of *Hmi* within the fibers as the above atoms are the coordination centers of the tetrapyrrole ring. Noteworthy is that the distribution of iron and chlorine atoms appears to be identical (hence, the EDX micrographs illustrating the presence of iron atoms is presented). In Figure 4, iron atoms are seen as orange dots. In the case of the neat PHB, neither iron nor chlorine atoms are seen.

As was mentioned above, PHB is a semicrystalline polymer, which is composed of amorphous and crystalline phases. The crystalline phase is presented by lamellae with folded chains [52]. Under certain conditions formation of spherulites is also allowed [53]. The amorphous phase exists in the rubbery state (glass transition temperature of PHB is 7 °C) The influence of technological methods was very great in the formation of the crystal structure of PHB, since it introduced boundary conditions in the process of formation of the supramolecular structure. Thus, the ES method promoted the formation of pass-through oriented macromolecules [54].

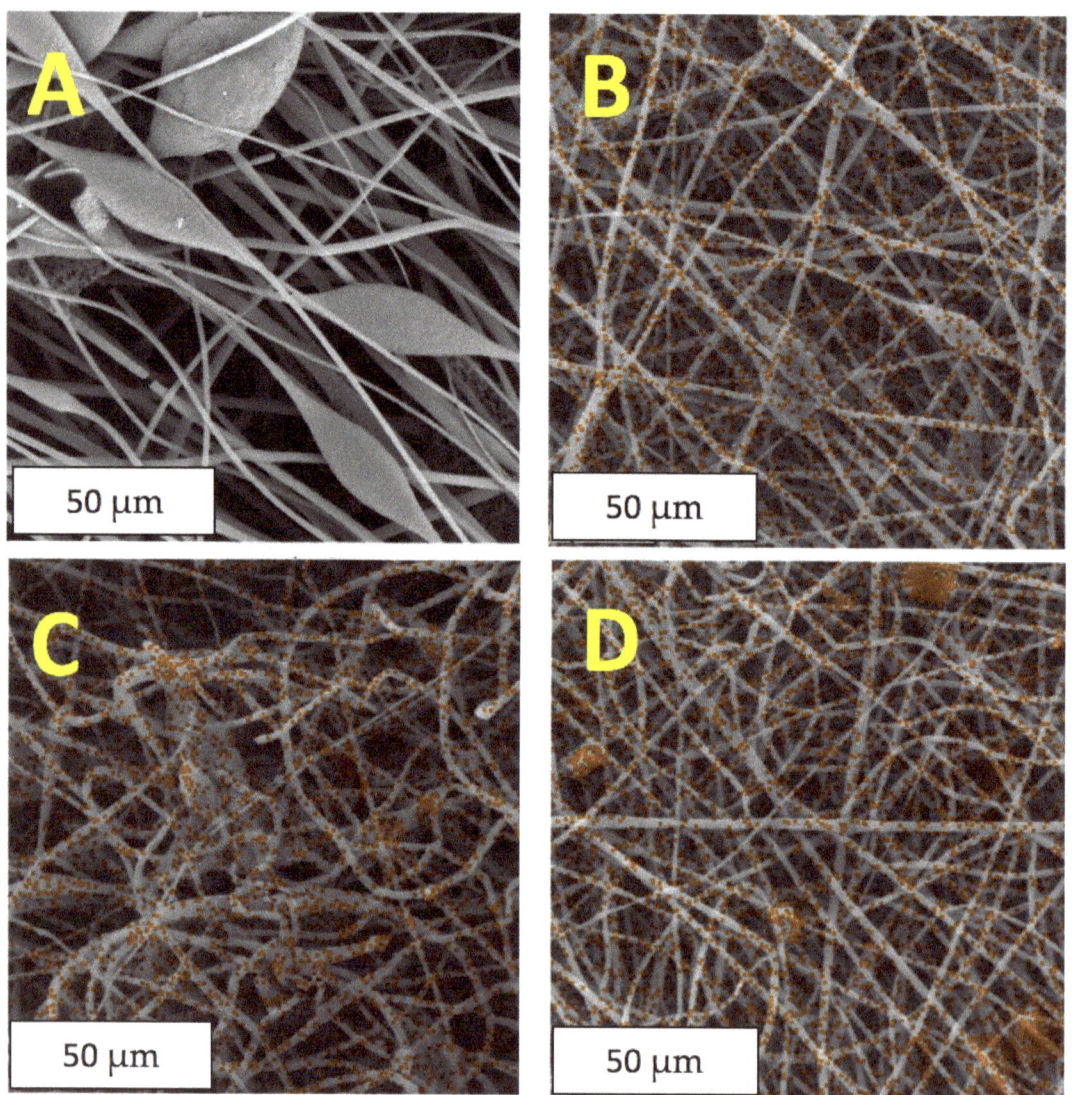

**Figure 4.** EDX elemental analysis of iron atoms in the *Hmi*/PHB samples with different content of hemin: (**A**) 0 wt.%, (**B**) 1 wt.%, (**C**) 3 wt.%, (**D**) 5 wt.%.

According to the DSC tests, the degree of crystallinity of PHB is ~60–65%. Hence, the content of the amorphous phase in the rubbery state is ~35–40% [55]. As follows from Figure 5, as the content of *Hmi* in the *Hmi*/PHB system increases to 5 wt.%, the degree of crystallinity decreases by 12%. Hence, the presence of *Hmi* has a certain effect on the crystallization of PHB from the solution during the ES process.

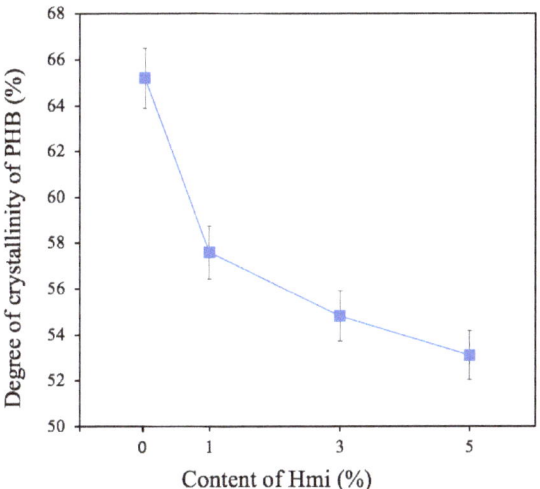

**Figure 5.** Degree of crystallinity versus the content of *Hmi* in the *Hmi*/PHB system.

Of special interest are the profiles of the melting DSC runs curve for the *Hmi*/PHB samples with the different content of *Hmi*. As the content of *Hmi* in the *Hmi*/PHB materials increases, the melting peak upon the first heating run (Figure 6A), becomes more symmetric, the melting temperature remains virtually unchanged, and the degree of crystallinity decreases. Hence, this evidence makes it possible to conclude that, in the presence of *Hmi*, crystallization of PHB proceeds at a slower rate and in a more uniform manner thus leading to the formation of a more regular and well-organized structure. Noteworthy is that, for the samples with a higher *Hmi* content, the low-temperature shoulder in the temperature interval of 150–160 °C becomes less pronounced, thus indicating the lower content of poorly organized crystallites and defects. On the second heating run, the DSC scans show double melting peaks, and this behavior is typical of PHB and corresponds to the melting-recrystallization-remelting process. The melting temperature of initial crystals is observed at 160 °C, whereas the melting temperature after recrystallization is about 174 °C.

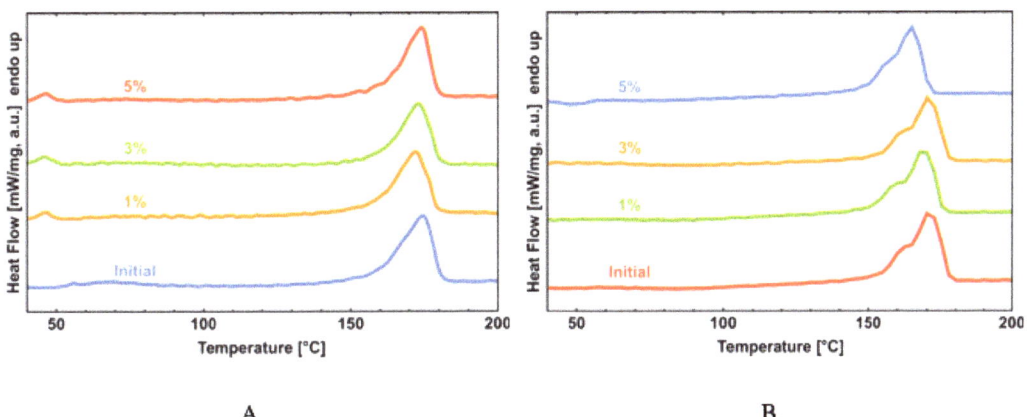

A.  B.

**Figure 6.** The DSC scans of Hmi/PHB with different content of Hmi: first heating run (**A**), second heating run (**B**).

The Hmi/PHB materials were studied by the EPR method to gain a deeper insight into the inner structure of the amorphous regions of PHB. Noteworthy is that the spin probe radical TEMPO is localized only within the amorphous phase of semicrystalline PHB whereas the crystalline regions contain no TEMPO. Hence, the analysis of the corresponding EPR spectra provides information concerning the mobility of spin radicals within the amorphous regions. Figure 7 shows the correlation times and concentration of TEMPO radicals plotted against the content of *Hmi* in the samples. As follows from Figure 7A, as the content of *Hmi* in the *Hmi*/PHB samples increases, the correlation time tends to decrease, and this decrease suggests the mobility of the TEMPO spin probe is reduced. The content of the spin probe (Figure 7B) appears to be independent of the *Hmi* content in the *Hmi*/PHB samples [54]. In other words, even when the concentration of the probe radical is the same, the mobility of the spin probes decreases.

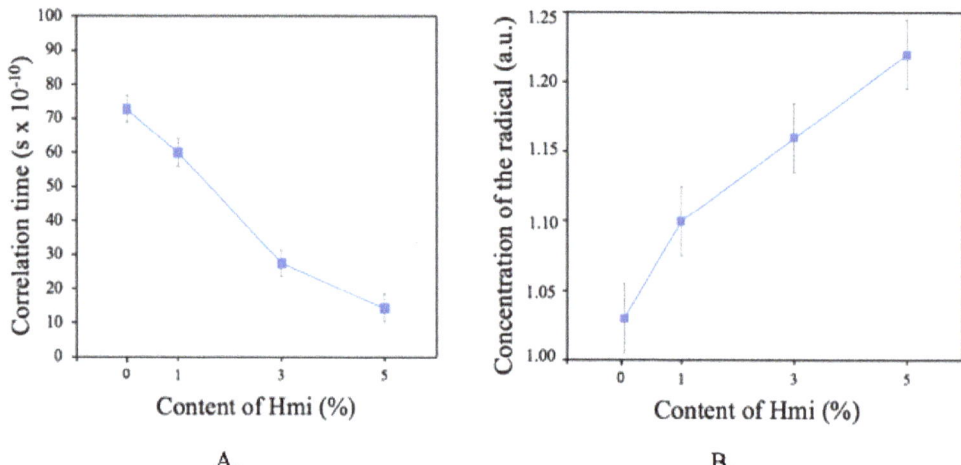

**Figure 7.** Correlation times (**A**) and concentration of TEMPO (**B**) versus the content of *Hmi* in the *Hmi*/PHB material.

Marked changes in supramolecular structure and morphology of the *Hmi*/PHB lead to changes in the physical-mechanical properties of the resultant materials (Table 2). For the *Hmi*/PHB containing 5 wt.% of *Hmi*, tensile strength increases up to 5.5 MPa, and this value is three times higher than that of the PHB nonwoven materials. Elongation at break is equal to 6% (1.7 times higher than that of the initial PHB).

**Table 2.** Mechanical characteristics of the *Hmi*/PHB samples with different content of *Hmi*.

| Content of *Hmi* % | Tensile Strength (Mean ± SD = 0.05, n = 10) MPa | Elongation at Break (Mean ± SD = 0.2, n = 10) % |
|---|---|---|
| 0 | 1.7 | 3.6 |
| 1 | 0.7 | 4.7 |
| 3 | 1.9 | 4.7 |
| 5 | 5.5 | 6.1 |

*3.2. Outdoor Exposure of Hemin/PHB Fibrous Materials*

The stability of the biodegradable *Hmi*/PHB materials containing 5 wt.% of *Hmi* was studied under outdoor exposure. The microscopic images of the test samples before and after weathering experiments are shown in Figure 8. In contrast to the PHB nonwovens containing porphyrins with metal ions when degradation within 60–90 days is complete, the *Hmi*/PHB materials remain nearly intact: the material retained its integrity and shape.

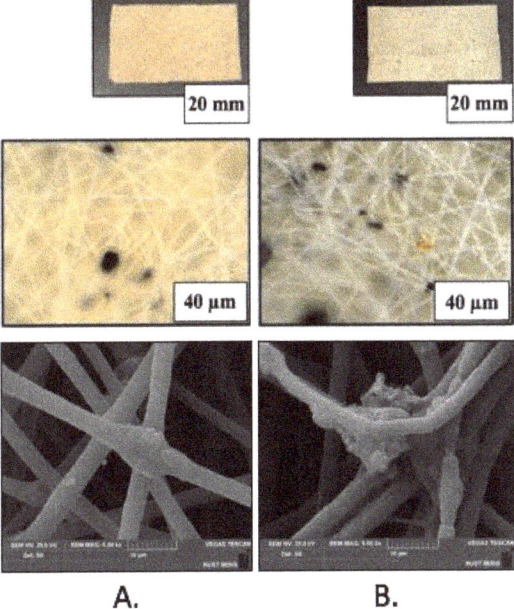

**A.**                    **B.**

**Figure 8.** Snapshots and SEM images of the *Hmi*/PHB samples before (**A**) and after (**B**) outdoor exposure.

The corresponding FTIR spectra of the initial and weathered samples are seen to be nearly the same and reveal no presence of the accumulated products of the degradation process (Figure 9).

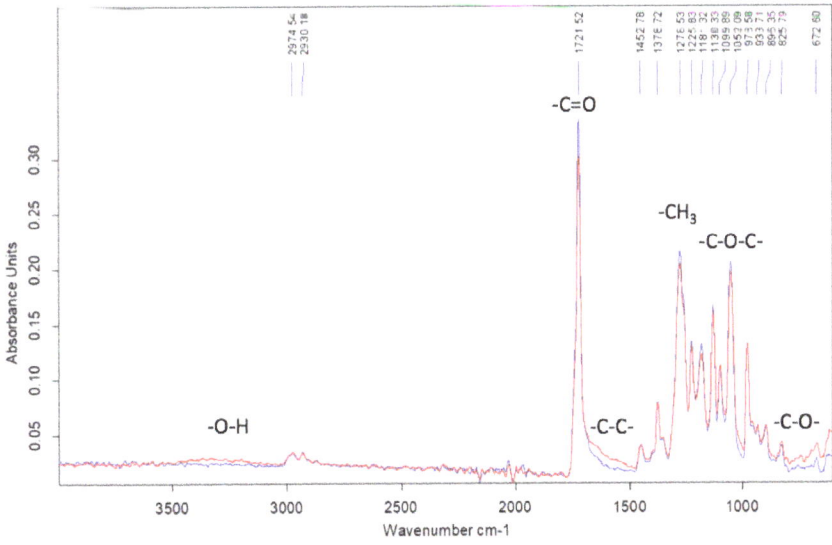

**Figure 9.** The FTIR spectra of pristine (blue line) and weathered (red line) *Hmi*/PHB samples.

The FTIR spectrum shows a slight decrease in the intensity of the peaks at 1721 cm$^{-1}$ (C=O group), 1278 cm$^{-1}$ (CH$_3$ group), and 1052 cm$^{-1}$, (C-O-C group). In the region of 900–700 and 1700–1500 cm$^{-1}$, noticeable changes are observed. This fact suggests the breakage of C-C bonds and the accumulation of OH groups.

Even though the weathered *Hmi*/PHB samples experience minor biodegradation, their supramolecular structure after weathering is changed. The results of the DSC analysis are presented in Figure 10. The shape of the melting peak is markedly changed, thus indicating the occurrence of certain processes during the exposure period.

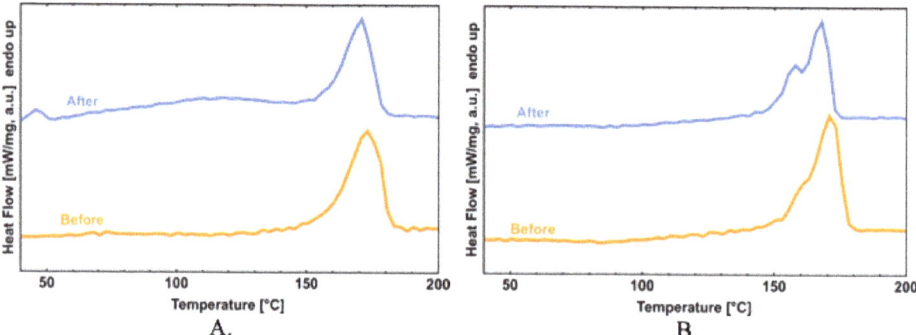

**Figure 10.** DSC scans of hemin/PHB (5 wt.%) nanocomposites before (yellow line) and after weathering (blue line): (**A**) first heating run and (**B**) second heating run.

The heat of fusion ($\Delta H$) and melting temperature ($T_m$) of intact and weathered PHB and *Hmi*/PHB (5 wt.%) nanocomposites are listed in Table 3.

**Table 3.** Thermophysical characteristics ($T_m$—the melting temperature, $\Delta H$—the heat of fusion) of weathered PHB and *Hmi*/PHB nanocomposites (90 days of exposure).

| Sample | Exposure Time Days | First Heating Run | | Second Heating Run | |
|---|---|---|---|---|---|
| | | $T_m$, °C | $\Delta H$, J/g | $T_m$, °C | $\Delta H$, J/g |
| *Hmi*/PHB (5 wt.%) | 0 | 174.5 | 88.7 | 171.0 | 87.9 |
| *Hmi*/PHB (5 wt.%) | 90 | 171.3 | 82.8 | 166.9 | 80.2 |
| PHB | 0 | 174.1 | 73.6 | 170.4 | 79.4 |
| PHB | 90 | 170.8 | 63.2 | 167.2 | 64.9 |

The corresponding stress-strain curves of the *Hmi*/PHB containing 5 wt.% of *Hmi* before and after weathering are shown in Figure 11.

As follows from Figure 11, upon weathering, the mechanical characteristics (tensile stress and elongation at break) are markedly deteriorated. Information on tensile strength and elongation at break is presented in Table 4.

Upon weathering, the tensile strength of the *Hmi*/PHB materials decreases by a factor of 2, and elongation at break decreases from 6.1 down to 4.2%. However, even after outdoor exposure, the *Hmi*/PHB materials (in contrast to the weathering of the neat PHB) preserve their stability and high mechanical properties. Reduced mechanical properties suggest the occurrence of certain irreversible destruction processes promoted by the photosensitizing activity of the tetrapyrrole complex of Hemin. However, the samples preserve good mechanical properties, and both strength and elongation are higher than those of the initial PHB. Slightly increased strength characteristics of PHB are known to be provided by oxidation and UV irradiation.

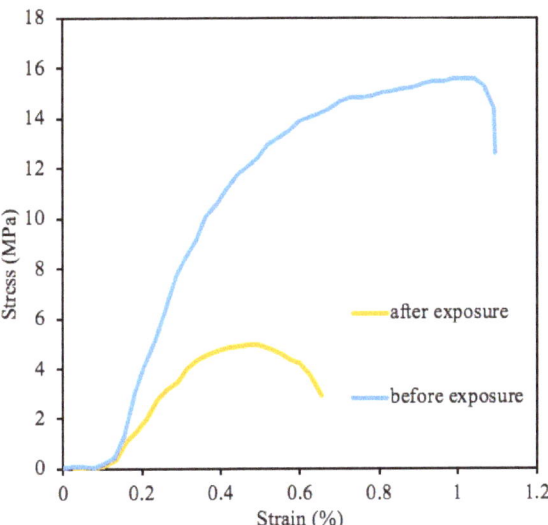

**Figure 11.** Strain-stress curves of the initial (blue line) and weathered (yellow line) samples.

**Table 4.** Mechanical characteristics of the neat and weathered *Hmi*/PHB materials (90 days of exposure).

| Content of *Hmi* % | Exposure Time Days | Tensile Strength (Mean ± SD = 0.05, *n* = 10) MPa | Elongation at Break (Mean ± SD = 0.2, *n* = 10) % |
|---|---|---|---|
| 5 | 0 | 5.5 | 6.1 |
| 5 | 90 | 1.9 | 4.5 |
| 0 | 0 | 1.7 | 3.6 |
| 0 | 90 | 2.7 | 4.2 |

## 4. Discussion

Nonwoven materials are characterized by three hierarchical levels of structural organization: macroscopic (the system in whole), mesoscopic (fiber contact area), and microscopic (structure of individual fibers) levels [56], and structural organization at all levels is controlled by the regime of electrospinning and composition of feed solution. For the *Hmi*/PHB composite, the introduction of hemin molecules to the feed mixture appears to exert a marked effect on the structural organization of *Hmi*/PHB at all levels.

At the stage of the electrospinning, the introduction of an even concentration of *Hmi* (1, 3, 5 wt.%) to the *Hmi*/PHB cocktail provides a sustainable regime of ES and leads to the formation of regular nonwoven fibrous material with a uniform distribution of *Hmi* within the amorphous phase between crystallites [51] as proved by the EDX method. The role of *Hmi* is likely to be associated with an increased electrical conductivity of the feed solution due to the presence of the metal ion as the coordination center in the *Hmi* molecule and the viscosity of the solution [57]. As a result, the electrospinning conditions are improved, and the performance of ES is enhanced. As the content of *Hmi* in the *Hmi*/PHB solution increases (from 1 to 5 wt.%), the fibrous material becomes more uniform, and the thickness of regular fibers without defects (smudges, spindles, etc.) decreases down to 1.77 μm. Noteworthy is that the fibers are well developed upon the Taylor cone formation stage [50]. In this case, due to the well-organized movement of a jet upon ES, the porosity of the *Hmi*/PHB material increases from 80 (pristine PHB) to 94% (*Hmi*/PHB containing 5 wt.%).

In other words, at the high content of *Hmi* (Table 1), the *Hmi*/PHB material is characterized by a high percentage of the air layer between well-cured fibers.

Concerning the effect of hemin on the morphology of the *Hmi*/PHB materials, the SEM observations (Figure 3) prove that the quality and uniform character of the resultant *Hmi*/PHB nonwoven materials are improved due to the presence of big-sized *Hmi* molecules in the intermolecular space of PHB; as a result, the content of amorphous regions increases, thus increasing the rate of solvent desorption. Evidently, the above changes are provided by the supramolecular structure of PHB [44]. Electrospinning is known to be accompanied by a high deformation rate of the feed polymer solution ($10^3$ s$^{-1}$) [51], thus governing the orientation of polymer chains [58]. Upon electrospinning, PHB macromolecules are organized as fibers containing alternating crystalline and amorphous regions [59]. Without the action of external forces, spherulites are formed.

As follows from the DSC scans (Figure 6b), with increasing the content of *Hmi* in the *Hmi*/PHB materials (from 0 to 5 wt.%), the melting peak becomes sharper and the onset of melting is shifted to higher temperatures. This evidence proves the formation of more regular crystallites with regular average dimensions and the absence of poorly organized small crystallites. However, the overall degree of crystallinity decreases (from 65% for neat PHB down to 53% for *Hmi*/PHB containing 5 wt.% of *Hmi*). Hemin is likely to serve as nucleation sites for the formation of uniform PHB crystallites; after crystallization, *Hmi* is accommodated within the amorphous phase of the *Hmi*/PHB materials. According to the EPR observations, as the content of *Hmi* in the *Hmi*/PHB samples increases, correlation time markedly decreases (Figure 8).

The above changes in the morphology of the *Hmi*/PHB materials are associated with the mechanical properties of the resultant materials. The addition of *Hmi* to the feed mixtures provides the formation of regular and uniform fibers and better-organized and well-cured nonwoven materials. The fibers contain no defects as bonds, spindles, smudges, and thinned regions, and their thickness is uniform along their length. Hence, the mechanical characteristics of the *Hmi*/PHB materials are improved.

The stability of the resultant *Hmi*/PHB materials was tested under the conditions of natural weathering (outdoor exposure). In general, weathering of PHB is accompanied by diverse processes, including oxidation, hydrolysis, photooxidation, stress-induced changes, etc., which lead to degradation and cleavage of the polymer chain to shorter fragments [60]. Degradation of biodegradable polymers strongly depends on many factors, including microbial activity, polymer composition, molecular weight, degree of crystallinity, temperature, humidity, pH, UV action, accessibility to oxygen and other oxidizing agents, etc. [61]. According to general knowledge, biodegradation is primarily provided by the biological activity of living organisms and leads to the decomposition of organic compounds into nontoxic products with lower molecular weights. Biodegradation of PHB under anaerobic conditions is accompanied by the formation of carbon dioxide ($CO_2$), water ($H_2O$), and methane; under aerobic conditions, $CO_2$ and $H_2O$ are formed [62–64]. Several bacteria and fungi, e.g., *Pseudomonas*, *Actinomadura*, *Penicillium Aspergillus* spp., *Microbispora*, *Saccharomonospora*, *Streptomyces*, *Thermoactinomyces*, and *Bacillusspp.*, can also degrade PHAs both aerobically and anaerobically. Degradation of ultrathin PHB fibers is known to proceed with a high rate as compared with that of PHB films [65] due to the three-dimensional structure and high surface area of nanofibers [65,66].

The results on the outdoor weathering of the *Hmi*/PHB nonwoven materials containing thin fibers seem to be of vital importance for the characterization of their stability and potential use in biomedical applications. As follows from Figure 9, even after outdoor exposure for 90 days, the weathered *Hmi*/PHB samples preserve their shape and integrity. Further tests reveal insignificant fiber cleavage and thinning. The FTIR analysis reveals that degradation proceeds without any accumulation of functional groups or toxic products. As follows from Table 4, this process leads to certain deterioration in the mechanical properties of the *Hmi*/PHB materials: tensile strength decreases by a factor of 2, and elongation at break decreases by a factor of 1.5 (Figure 11). This evidence can be explained by the degradation-induced changes in the integrity of fibers and supramolecular structure of the *Hmi*/PHB materials due to oxidative and photooxidative processes in the air under the action of the UV light. Earlier, this phenomenon was observed for ozone-induced oxidation of PHB [67].

The DSC data [68] were analyzed for the detailed description of the structural evolution in the crystalline phase as evidenced by marked changes in physicomechanical properties.

Within 90 days of exposure, the supramolecular structure (including the crystalline phase) experiences marked changes in the action of environmental factors (water, wind, UV, temperature changes). The slight decrease in the melting temperature of the weathered samples suggests the formation of poorly organized crystallites and a certain decrease in the molecular weight of the samples [69,70]. However, in general, degradation of the *Hmi*/PHB materials is retarded as compared with the neat PHB. Seemingly, the exposure to solar radiation leads to the initiation of oxidative processes on the surface of individual fibers, leading to minor changes in the crystalline structure as evidenced by a well-pronounced shoulder and double melting peaks which are indicative of the formation of imperfect crystallites. Noteworthy is that the overall degree of crystallinity remains unchanged.

Let us also mention the action of *Hmi* as an efficient antibacterial agent with respect to bacteria and fungi [34,71–73]. In the presence of *Hmi*, intensive generation of hydroxyl radical leads to oxidation of cellular compartments, cell wall rupturing, and membrane leakage, thus providing bacterial apoptosis. As a result, the amorphous phase of the *Hmi*/PHB composite materials, which is the most susceptible to biodegradation, appears to be stabilized. Hence, the resultant *Hmi*/PHB composites with high biocompatibility, good mechanical properties, and retarded biodegradation can be used as effective biomedical materials and scaffolds for regenerative medicine and tissue engineering.

## 5. Conclusions

This work addresses the preparation of bioinspired *Hmi*/PHB composite materials with high mechanical properties and retarded biodegradation. The addition of *Hmi* to the feed PHB solution (above 3 wt.%) provides a marked improvement in the performance of the electrospinning process and leads to the formation of well-organized and uniform nonwoven materials with high porosity and good mechanical properties.

The work investigated and described the regularities of changes in the PHB-hemin system during outdoor exposure. The obtained results revealed a deeper understanding of the structural organization of PHB macromolecules and their response to oxidative processes. In the case of the commercial launch of the material, practical recommendations for the use of the PHB-hemin system can be carried out. This material retains its stability within 90 days. The *Hmi*/PHB materials can be recommended for their use as biomedical materials for wound disinfection and accelerated healing, and as scaffolds in regenerative medicine and tissue engineering, and this challenging task requires further detailed studies.

**Supplementary Materials:** The following supporting information can be downloaded at: https://www.mdpi.com/article/10.3390/polym14224878/s1, Figure S1: Photo and schematic view of the single-capillary laboratory unit for the electrospinning process; Figure S2: Histogram the fibers' size distribution of PHB and *Hmi*/PHB nonwoven composites.

**Author Contributions:** Conceptualization, O.V.A.; methodology and experimental design, P.M.T.; validation, O.V.A.; investigation A.K.Z., P.M.T., P.V.I., A.A.O., A.Y.Y. and I.A.V.; DSC tests, P.M.T.; writing—original draft preparation, P.M.T.; writing—review and editing, O.V.A.; supervision, O.V.A.; project administration, A.K.Z.; funding acquisition, P.M.T. All authors have read and agreed to the published version of the manuscript.

**Funding:** This research was funded by the Russian Science Foundation (Agreement No. 22-73-00038).

**Institutional Review Board Statement:** Not applicable.

**Informed Consent Statement:** Not applicable.

**Acknowledgments:** The study was carried out using scientific equipment of the Center of Shared Usage «New Materials and Technologies» of Emanuel Institute of Biochemical Physics and the Common Use Centre of Plekhanov Russian University of Economics.

**Conflicts of Interest:** The authors declare no conflict of interest.

# References

1. Poltronieri, P.; Kumar, P. Polyhydroxyalcanoates (PHAs) in Industrial Applications. In *Handbook of Ecomaterials*; Springer: Cham, Switzerland, 2017; pp. 1–30. [CrossRef]
2. Rajan, K.P.; Thomas, S.P.; Gopanna, A.; Chavali, M. Polyhydroxybutyrate (PHB): A Standout Biopolymer for Environmental Sustainability. In *Handbook of Ecomaterials*; Springer International Publishing AG: Cham, Switzerland, 2017; pp. 1–23. [CrossRef]
3. Kostopoulos, V.; Kotrotsos, A.; Fouriki, K.; Kalarakis, A.; Portan, D. Fabrication and Characterization of Polyetherimide Electrospun Scaffolds Modified with Graphene Nano-Platelets and Hydroxyapatite Nano-Particles. *Int. J. Mol. Sci.* **2020**, *21*, 583. [CrossRef] [PubMed]
4. Dror, Y.; Salalha, W.; Khalfin, R.L.; Cohen, Y.; Yarin, A.L.; Zussman, E. Carbon nanotubes embeded in oriented polymer nanofibers by electrospinning. *Langmuir* **2003**, *19*, 7012–7020. [CrossRef]
5. Moydeen, M.A.; Syed, A.M.; Padusha, E.F.; Aboelfetoh, S.S.; Al-Deyab, M.H. Fabrication of electrospun poly(vinyl alcohol)/dextran nanofibers via emulsion process as drug delivery system: Kinetics and in vitro release study. *Int. J. of Biol. Macromol.* **2018**, *116*, 1250–1259. [CrossRef] [PubMed]
6. Lund, A.; van der Velden, N.; Persson, N.-K.; Hamedi, M.; Müller, C. Electrically conducting fibres for e-textiles: An open playground for conjugated polymers and carbon nanomaterials. *Mater. Sci. Eng. R Rep.* **2018**, *126*, 1–29. [CrossRef]
7. Bailey, F.; Malinski, T.; Kiechle, F. Carbon-fiber ultramicroelectrodes modified with conductive polymeric tetrakis(3-methoxy-4-hydroxyphenyl)porphyrin for determination of nickel in single biological cells. *Anal. Chem.* **1991**, *63*, 395–398. [CrossRef]
8. Avossa, J.; Paolesse, R.; Di Natale, C.; Zampetti, E.; Bertoni, G.; De Cesare, F.; Scarascia-Mugnozza, G.; Macagnano, A. Electrospinning of polystyrene/polyhydroxybutyrate nanofibers doped with porphyrin and graphene for chemiresistor gas sensors. *Polymers* **2019**, *9*, 280. [CrossRef]
9. Tabassum, R.; Kant, R. Recent trends in surface plasmon resonance based fiber–optic gas sensors utilizing metal oxides and carbon nanomaterials as functional entities. *Sens. Actuators B Chem.* **2020**, *310*, 127813. [CrossRef]
10. Scheicher, S.R.; Kainz, B.; Köstler, S.; Suppan, M.; Bizzarri, A.; Pum, D.; Sleytr, U.; Ribitsch, V. Optical oxygen sensors based on Pt(II) porphyrin dye immobilized on S-layer protein matrices. *Biosens. Bioelectron.* **2009**, *25*, 797–802. [CrossRef]
11. Laskowska, M.; Kityk, L.; Pastukh, O.; Dulski, M.; Zubko, M.; Jedryka, J.; Laskowski, E. Nanocomposite for photonics—Nickel pyrophosphate nanocrystals synthesised in silica nanoreactors. *Microporous Mesoporous Mater.* **2020**, *306*, 110435. [CrossRef]
12. Biswas, S.; Ahn, H.-Y.; Bondar, M.V.; Belfield, K.D. Two-photon absorption enhancement of polymer-templated porphyrin-based J-aggregates. *Langmuir* **2012**, *28*, 1515–1522. [CrossRef] [PubMed]
13. Chen, Z.; Mai, B.; Tan, H.; Chen, X. The effect of thermally developed SiC@SiO$_2$ core-shell structured nanoparticles on the mechanical, thermal and UV-shielding properties of polyimide composites. *Compos. Commun.* **2018**, *10*, 194–204. [CrossRef]
14. Suo, Z.; Chen, J.; Hou, X.; Hu, Z.; Xing, F.; Feng, L. Growing prospects of DNA nanomaterials in novel biomedical applications. *RSC Adv.* **2019**, *9*, 16479–16491. [CrossRef] [PubMed]
15. Ghosal, K.; Agatemor, C.; Špitálsky, Z.; Thomas, S.; Kny, E. Electrospinning tissue engineering and wound dressing scaffolds from polymer–titanium dioxide nanocomposites. *Chem. Eng. J.* **2018**, *358*, 1262–1278. [CrossRef]
16. Wu, J.; Li, S.; Wei, H. Integrated nanozymes: Facile preparation and biomedical applications. *Chem. Commun.* **2018**, *54*, 6520–6530. [CrossRef]
17. Ruthard, C.; Schmidt, M.; Gröhn, F. Porphyrin-polymer networks, worms, and nanorods: pH-triggerable hierarchical self-assembly. *Macromol. Rapid. Commun.* **2011**, *32*, 706–711. [CrossRef]
18. Sessler, J.L.; Tomat, E. Transition-metal complexes of expanded porphyrins. *Acc. Chem. Res.* **2007**, *40*, 371–379. [CrossRef]
19. Sessler, J.; Miller, R. Texaphyrins: New drugs with diverse clinical applications in radiation and photodynamic therapy. *Biochem. Pharmacol.* **2000**, *59*, 733–739. [CrossRef]
20. Tsolekile, N.; Nelana, S.; Oluwafemi, O.S. Porphyrin as diagnostic and therapeutic agent. *Molecules* **2019**, *24*, 2669. [CrossRef]
21. Habermeyer, B.; Guilard, R. Some activities of Porphy Chem illustrated by the applications of porphyrinoids in PDT, PIT and PDI. *Photochem. Photobiol. Sci.* **2018**, *17*, 1675–1690. [CrossRef]
22. Zhang, L.; Yang, R.; Yu, H.; Xu, Z.; Kang, Y.; Cui, H.; Xue, P. MnO$_2$-capped silk fibroin (SF) nanoparticles with chlorin e6 (Ce6) encapsulation for augmented photo-driven therapy by modulating the tumor microenvironment. *J. Mater. Chem. B* **2021**, *9*, 3677–3688. [CrossRef] [PubMed]
23. Alsharabasy, A.M.; Pandit, A.; Farràs, P. Recent advances in the design and sensing applications of hemin/coordination polymer-based nanocomposites. *Adv. Mater.* **2020**, *33*, 2003883. [CrossRef] [PubMed]
24. Sasaki, D.; Watanabe, T.F.; Eady, R.R.; Garratt, R.C.; Antonyuk, S.V.; Hasnain, S.S. Reverse protein engineering of a novel 4-domain copper nitrite reductase reveals functional regulation by protein-protein interaction. *FEBS J.* **2021**, *288*, 262–280. [CrossRef]
25. Zhang, Y.; Xu, C.; Li, B. Self-assembly of hemin on carbon nanotube as highly active peroxidase mimetic and its application for biosensing. *RSC Adv.* **2013**, *3*, 6044. [CrossRef]

26. Dell'Acqua, S.; Massardi, E.; Monzani, E.; Di Natale, G.; Rizzarelli, E.; Casella, L. Interaction between hemin and prion peptides: Binding, oxidative reactivity and aggregation. *Int. J. Mol. Sci.* **2020**, *21*, 7553. [CrossRef] [PubMed]
27. Zozulia, O.; Korendovych, I.V. Semi-rationally designed short peptides self-assemble and bind hemin to promote cyclopropanation. *Angew. Chem. Int. Ed.* **2020**, *59*, 8108–8112. [CrossRef]
28. Qu, R.; Shen, L.; Chai, Z.; Jing, C.; Zhang, Y.; An, Y.; Shi, L. Hemin-block copolymer micelle as an artificial peroxidase and its applications in chromogenic detection and biocatalysis. *ACS Appl. Mater. Interfaces* **2014**, *6*, 19207–19216. [CrossRef]
29. Wang, J.; Cao, Y.; Chen, G.; Li, G. Regulation of thrombin activity with a bifunctional aptamer and hemin: Development of a new anticoagulant and antidote pair. *Chembiochem* **2009**, *10*, 2171–2176. [CrossRef]
30. Zhao, Y.; Zhang, L.; Wei, W.; Li, Y.; Liu, A.; Zhang, Y.; Liu, S. Effect of annealing temperature and element composition of titanium dioxide/graphene/hemin catalysts for oxygen reduction reaction. *RSC Adv.* **2015**, *5*, 82879–82886. [CrossRef]
31. Nitzan, Y.; Ladan, H.; Gozansky, S.; Malik, Z. Characterization of hemin antibacterial action on Staphylococcus aureus. *FEMS Microbiol. Lett.* **1987**, *48*, 401–406. [CrossRef]
32. Sedaghat, S.; Shamspur, T.; Mohamadi, M.; Mostafavi, A. Extraction and preconcentration of hemin from human blood serum and breast cancer supernatant. *J. Sep. Sci.* **2015**, *38*, 4286–4291. [CrossRef] [PubMed]
33. Sun, L.; Song, L.; Zhang, X.; Zhou, R.; Yin, J.; Luan, S. Poly(γ-glutamic acid)-based electrospun nanofibrous mats with photodynamic therapy for effectively combating wound infection. *Mater. Sci. Eng. C* **2020**, *113*, 110936. [CrossRef] [PubMed]
34. Min, T.; Sun, X.; Zhou, L.; Du, H.; Zhu, Z.; Wen, Y. Electrospun pullulan/PVA nanofibers integrated with thymol-loaded porphyrin metal-organic framework for antibacterial food packaging. *Carbohydr. Polym.* **2021**, *270*, 118391. [CrossRef] [PubMed]
35. Gangemi, C.M.A.; Iudici, M.; Spitaleri, L.; Randazzo, R.; Gaeta, M.; D'Urso, A.; Fragalà, M.E. Polyethersulfone mats functionalized with porphyrin for removal of para-nitroaniline from aqueous solution. *Molecules* **2019**, *24*, 3344. [CrossRef] [PubMed]
36. Costa, M.S.; Fangueiro, R.; Ferreira, P.D. Drug Delivery Systems for Photodynamic Therapy: The Potentiality and Versatility of Electrospun Nanofibers. *Macromol. Biosci.* **2022**, *22*, e2100512. [CrossRef]
37. Lv, D.; Wang, R.; Tang, G.; Mou, Z.; Lei, J.; Han, J.; Huang, C. Eco-friendly Electrospun Membranes Loaded with Visible-light Response Nano-particles for Multifunctional usages: High-efficient Air Filtration, Dye Scavenger and Bactericide. *ACS Appl. Mater. Interfaces* **2019**, *11*, 12880–12889. [CrossRef]
38. Lu, T.; Cui, J.; Qu, Q.; Wang, Y.; Zhang, J.; Xiong, R.; Huang, C. Multistructured Electrospun Nanofibers for Air Filtration: A Review. *ACS Appl. Mater. Interfaces* **2021**, *13*, 23293–23313. [CrossRef]
39. Nakayama, D.; Wu, F.; Mohanty, A.K.; Hirai, S.; Misra, M. Biodegradable Composites Developed from PBAT/PLA Binary Blends and Silk Powder: Compatibilization and Performance Evaluation. *ACS Omega* **2018**, *3*, 12412–12421. [CrossRef]
40. Nofar, M.; Salehiyan, R.; Ciftci, U.; Jalali, A.; Durmuş, A. Ductility improvements of PLA-based binary and ternary blends with controlled morphology using PBAT, PBSA, and nanoclay. *Compos. B. Eng.* **2019**, *182*, 107661. [CrossRef]
41. Huang, M.-H.; Li, S.; Hutmacher, D.W.; Coudane, J.; Vert, M. Degradation characteristics of poly(ε-caprolactone)-based copolymers and blends. *J. Appl. Polym. Sci* **2016**, *102*, 1681–1687. [CrossRef]
42. Dehant, I. *Infrared Spectroscopy Of Polymers*; Chemistry: Moscow, Russia, 1976.
43. Tarasevich, B.N. IR spectra of general classes of organic compounds. In *Reference Material*; Publishing House of MSU: Moscow, Russia, 2012.
44. Bellamy, L. *The Infra-Red Spectra of Complex Molecules*; Springer: Amsterdam, The Netherlands, 1975.
45. Vyazovkin, S.; Koga, N.; Schick, C. *Handbook of Thermal Analysis and Calorimetry, Applications to Polymers and Plastics*; Elsevier: London, UK, 2002.
46. Zhu, G.; Kremenakova, D.; Wang, Y.; Militky, J. Air permeability of polyester nonwoven fabrics. *Autex Res. J.* **2015**, *15*, 8–12. [CrossRef]
47. Scandola, M.; Focarete, M.L.; Adamus, G.; Sikorska, W.; Baranowska, I.; Świerczek, S.; Gnatowski, M.; Kowalczuk, M.; Jedliński, Z. Polymer blends of natural poly(3-hydroxybutyrate-co-3-hydroxyvalerate) and a synthetic atactic poly(3-hydroxybutyrate). Characterization and biodegradation studies. *Macromolecules* **1997**, *30*, 2568–2574. [CrossRef]
48. Sezer, D.; Freed, J.H.; Roux, B. Simulating electron spin resonance spectra of nitroxide spin labels from molecular dynamics and stochastic trajectories. *J. Chem. Phys.* **2008**, *128*, 165106. [CrossRef] [PubMed]
49. Domaschke, S.; Zündel, M.; Mazza, E.; Ehret, A.E. A 3D computational model of electrospun networks and its application to inform a reduced modelling approach. *Int. J. Solids Struct.* **2019**, *178*, 76–89. [CrossRef]
50. Katsogiannis, K.A.G.; Vladisavljević, G.T.; Georgiadou, S. Porous electrospun polycaprolactone (PCL) fibres by phase separation. *Eur. Polym. J.* **2015**, *69*, 284–295. [CrossRef]
51. Reneker, D.H.; Yarian, A.L.; Zussman, E.; Xu, H. Electrospinning of nanofibers from polymer solutions and melts. *Adv. Appl. Mech.* **2007**, *41*, 43–195.
52. Ding, G.; Liu, J. Morphological varieties and kinetic behaviors of poly(3-hydroxybutyrate) (PHB) spherulites crystallized isothermally from thin melt film. *Colloid Polym. Sci.* **2013**, *291*, 1547–1554. [CrossRef]
53. Zhao, L.; Qu, R.; Li, A.; Ma, R.; Shi, L. Cooperative self-assembly of porphyrins with polymers possessing bio-active functions. *Chem. Comm.* **2016**, *52*, 13543–13555. [CrossRef]
54. Tyubaeva, P.; Varyan, I.; Krivandin, A.; Shatalova, O.; Karpova, S.; Lobanov, A.; Olkhov, A.; Popov, A. The Comparison of Advanced Electrospun Materials Based on Poly(-3-hydroxybutyrate) with Natural and Synthetic Additives. *J. Funct. Biomater.* **2022**, *13*, 23. [CrossRef]

55. Arai, T.; Tanaka, M.; Kawakami, H. Porphyrin-containing electrospun nanofibers: Positional control of porphyrin molecules in nanofibers and their catalytic application. *ACS Appl. Mater. Interfaces* **2012**, *4*, 5453–5457. [CrossRef]
56. Syerko, E.; Comas-Cardona, S.; Binetruy, C. Models of mechanical properties/behavior of dry fibrous materials at various scales in bending and tension: A review. *Compos. Part A Appl. Sci. Manuf.* **2012**, *43*, 1365–1388. [CrossRef]
57. Baumgarten, P.K. Electrostatic spinning of acrylic microfibers. *J. Colloid Interface Sci.* **1971**, *36*, 71–79. [CrossRef]
58. Allais, M.; Mailley, D.; Hébraud, P.; Ihiawakrim, D.; Ball, V.; Meyer, F.; Hébraud, A.; Schlatter, G. Polymer-free electrospinning of tannic acid and cross-linking in water for hybrid supramolecular nanofibres. *Nanoscale* **2018**, *10*, 9164–9173. [CrossRef]
59. Hannay, N.B. *Treatise on Solid State Chemistry, Crystalline and Noncrystalline Solids*; Plenum Press: New York, NY, USA, 1976. [CrossRef]
60. Kumar, S.; Pal, S.; Ray, S. Study of microbes having potentiality for biodegradation of plastics. *Environ. Sci. Pollut. Res.* **2013**, *20*, 4339–4355. [CrossRef]
61. Boopathy, R. Factors limiting bioremediation technologies. *Bioresour. Technol.* **2000**, *74*, 63–67. [CrossRef]
62. Wang, Z. Biodegradation of polyhydroxybutyrate film by Pseudomonas mendocinaDS04-T. *Polym. Plast. Technol. Eng.* **2013**, *52*, 195–199. [CrossRef]
63. Bonartseva, G.; Myshkina, V.; Nikolaeva, D.; Kevbrina, M.; Kallistova, A.; Gerasin, V.; Iordanskii, A.; Nozhevnikova, A. Aerobic and anaerobic microbial degradation of poly-β-hydroxybutyrate produced by Azotobacter chroococcum. *Appl. Biochem. Biotechnol.* **2003**, *109*, 285–301. [CrossRef]
64. Vert, M.; Doi, Y.; Hellwich, K.-H.; Hess, M.; Hodge, P.; Kubisa, P.; Rinaudo, M.; Schué, F. Terminology for biorelated polymers and applications (IUPAC Recommendations 2012). *Pure Appl. Chem.* **2012**, *84*, 377–410. [CrossRef]
65. Altaee, N.; El-Hiti, G.A.; Fahdil, A.; Sudesh, K.; Yousif, E. Biodegradation of different formulations of polyhydroxybutyrate films in soil. *Springer Plus* **2016**, *5*, 762. [CrossRef]
66. Tokiwa, Y.; Calabia, B.; Ugwu, C.; Aiba, S. Biodegradability of plastics. *Int. J. Mol. Sci.* **2009**, *10*, 3722–3742. [CrossRef]
67. Tyubaeva, P.; Zykova, A.; Podmasteriev, V.; Olkhov, A.; Popov, A.; Iordanskii, A. The investigation of the structure and properties of ozone-sterilized nonwoven biopolymer materials for medical applications. *Polymers* **2021**, *13*, 1268. [CrossRef] [PubMed]
68. Mosnáčková, K.; Danko, M.; Šišková, A.; Falco, L.; Janigová, I.; Chmela, Š.; Vanovčanová, Z.; Omaníková, L.; Chodáka, I.; Mosnáček, J. Complex study of the physical properties of a poly(lactic acid)/poly(3-hydroxybutyrate) blend and its carbon black composite during various outdoor and laboratory ageing conditions. *RSC Adv.* **2017**, *7*, 47132–47142. [CrossRef]
69. Chen, Y.; Chou, I.-N.; Tsai, Y.; Wu, H.-S. Thermal degradation of poly(3-hydroxybutyrate) and poly(3-hydroxybutyrate-co-3-hydroxyvalerate) in drying treatment. *J. Appl. Polym. Sci.* **2013**, *130*, 3659–3667. [CrossRef]
70. Haslböck, M.; Klotz, M.; Steiner, L.; Sperl, J.; Sieber, V.; Zollfrank, C.; Van Opdenbosch, D. Structures of mixed-tacticity polyhydroxybutyrates. *Macromolecules* **2018**, *51*, 5001–5010. [CrossRef]
71. Ladan, H.; Nitzan, Y.; Malik, Z. The antibacterial activity of haemin compared with cobalt, zinc and magnesium protoporphyrin and its effect on pottassium loss and ultrastructure of Staphylococcus aureus. *FEMS Microbiol. Lett.* **1993**, *112*, 173–177. [CrossRef]
72. Stojiljkovic, I.; Evavold, B.D.; Kumar, V. Antimicrobial properties of porphyrins. *Expert Opin. Investig. Drugs* **2001**, *10*, 309–320. [CrossRef] [PubMed]
73. Le Guern, F.; Ouk, T.S.; Yerzhan, I.; Nurlykyz, Y.; Arnoux, P.; Frochot, C.; Leroy-Lhez, S.; Sol, V. Photophysical and Bactericidal Properties of Pyridinium and Imidazolium Porphyrins for Photodynamic Antimicrobial Chemotherapy. *Molecules* **2021**, *26*, 1122. [CrossRef] [PubMed]

Article

# Emulsion Stabilization by Cationic Lignin Surfactants Derived from Bioethanol Production and Kraft Pulping Processes

Avido Yuliestyan [1], Pedro Partal [2,*], Francisco J. Navarro [2], Raquel Martín-Sampedro [3], David Ibarra [3] and María E. Eugenio [3]

[1] Department of Chemical Engineering, Faculty of Industrial Technology, UPN Veteran Yogyakarta, Jalan SWK 104, Yogyakarta 55283, Indonesia; avido.yuliestyan@upnyk.ac.id
[2] Pro2TecS-Chemical Process and Product Technology Research Centre, Department of Chemical Engineeing, ETSI, Campus de "El Carmen", University of Huelva, 21071 Huelva, Spain; frando@uhu.es
[3] Forest Research Center (INIA, CSIC), Ctra. de la Coruña km 7.5, 28040 Madrid, Spain; raquel.martin@inia.csic.es (R.M.-S.); ibarra.david@inia.csic.es (D.I.); mariaeugenia@inia.csic.es (M.E.E.)
* Correspondence: partal@uhu.es

**Abstract:** Oil-in-water bitumen emulsions stabilized by biobased surfactants such as lignin are in line with the current sustainable approaches of the asphalt industry involving bitumen emulsions for reduced temperature asphalt technologies. With this aim, three lignins, derived from the kraft pulping and bioethanol industries, were chemically modified via the Mannich reaction to be used as cationic emulsifiers. A comprehensive chemical characterization was conducted on raw lignin-rich products, showing that the kraft sample presents a higher lignin concentration and lower molecular weight. Instead, bioethanol-derived samples, with characteristics of non-woody lignins, present a high concentration of carbohydrate residues and ashes. Lignin amination was performed at pH = 10 and 13, using tetraethylene pentamine and formaldehyde as reagents at three different stoichiometric molar ratios. The emulsification ability of such cationic surfactants was firstly studied on prototype silicone oil-in-water emulsions, attending to their droplet size distribution and viscous behavior. Among the synthetized surfactants, cationic kraft lignin has shown the best emulsification performance, being used for the development of bitumen emulsions. In this regard, cationic kraft lignin has successfully stabilized oil-in-water emulsions containing 60% bitumen using small surfactant concentrations, between 0.25 and 0.75%, which was obtained at pH = 13 and reagent molar ratios between 1/7/7 and 1/28/28 (lignin/tetraethylene pentamine/formaldehyde).

**Keywords:** lignin; amination; emulsion; bitumen; rheology; microstructure

**Citation:** Yuliestyan, A.; Partal, P.; Navarro, F.J.; Martín-Sampedro, R.; Ibarra, D.; Eugenio, M.E. Emulsion Stabilization by Cationic Lignin Surfactants Derived from Bioethanol Production and Kraft Pulping Processes. *Polymers* **2022**, *14*, 2879. https://doi.org/10.3390/polym14142879

Academic Editor: Eduardo Guzmán

Received: 27 May 2022
Accepted: 12 July 2022
Published: 15 July 2022

**Publisher's Note:** MDPI stays neutral with regard to jurisdictional claims in published maps and institutional affiliations.

## 1. Introduction

Lignin is after cellulose the second most abundant renewable polymer [1,2], being mainly synthetized by enzymatic polymerization of the following monomers: p-coumaryl alcohol, coniferyl alcohol, and synapyl alcohol. Each of these monolignols gives rise to different types of lignin units, which are called p-hydroxyphenyl (H), guaiacyl (G), and syringyl (S), respectively [3]. It is known that other monomers such as coniferaldehyde, acylated monolignols, etc. could also act as precursors of lignin [3]. The final polymeric structure of lignin is a heterogeneous complex macromolecule with different types of bonds and a large variety of functional groups that depend on the biomass source [4]. Accordingly, hardwood would contain approximately 20–25% of lignin, which is mostly composed of G and S units and traces of H units. On the other hand, softwood with around 20–35% of lignin mostly contains G units and small amount of H units [5,6]. Finally, non-woody species have lower lignin content (9–20%) of HGS type, which is characterized by a high proportion of H units [7]. Moreover, along with their source, the method used to generate the different lignins strongly affects their characteristics and, consequently, industrial applications [8,9].

Nowadays, pulp and paper industries, using processes such as kraft and sulfite pulping, generate a high amount of residual lignin dissolved in the black liquor, yielding kraft lignin and lignosulfonates [10]. Around 50–70 million tons/year of lignin are generated by this industry, mainly from kraft mills, since it is the most extended pulping technology to produce cellulose pulp [11,12]. On the other hand, the residues generated in lignocellulosic biorefineries that produce bioethanol are also enriched in lignin [8]. Within these biorefineries, the residual lignins come either from the solid residues generated after saccharification/fermentation processes (lignin of lower quality) or from the liquid fractions generated in the pre-treatments carried out on the raw biomass (high purity lignin) [13]. If these lignins are also considered, an increase of up to 225 million tons/year of residual lignins is forecasted by 2030 [14].

Most of such lignin side streams are normally burned, due to their high calorific value, to obtain energy that is partially used in the same production plant. However, this process generates an excess of energy, making more interesting the valorization of the residual lignin into high value-added products that guarantee the profitability of these plants [15]. Interestingly, a wide range of products can be obtained from them with application in many different industries, given the interesting properties that these residual lignins may exhibit (e.g., biodegradability, antioxidant and antimicrobial capacity, high reinforcing ability, surfactant properties) and the possibility of chemical modifications due to the high amount of chemicals sites available in the lignin structure [16].

Among promising alternatives, lignin-derived products could find application in the asphalt industry that has demonstrated a strong commitment to the development of more sustainable technologies and materials. In this regard, reduced temperature asphalt technologies, involving cold or half warm mix asphalt, may lead to a remarkable reduction in energy consumption (up to 7.5 times lower) and emissions (up to 7 times less kg equivalent $CO_2$/ton product) [17]. For such technologies, the use of a bituminous emulsion allows bitumen mixing with aggregate and further compaction operations to be performed at lower temperatures (in the range 60–130 °C) compared with those used for hot mix asphalt, which is typically above 165 °C [17–19]. Moreover, the development of oil-in-water bitumen emulsions stabilized by a biobased surfactant such as lignin would be clearly in line with the above commented sustainability approach.

However, markets and similar product specifications demand positively charged cationic bitumen emulsions [20], given that most common aggregates are of acidic type (i.e., siliceous-based derivatives) and exhibit a negative charge on their surfaces [21]. Due to the absence of protonatable functional groups, raw lignins cannot be used as cationic emulsifiers requiring their amination through the well-known Mannich reaction, which involves formaldehyde and amines as reagents, and this may be carried out under alkaline conditions [22,23]. The resultant modified lignins become positively charged and soluble under an acidic aqueous phase of the emulsion. These charges, when used in cationic bitumen emulsions for asphalt, are expected to stabilize the bitumen–water interface and, more interestingly, promote binder interaction and adhesion to a negatively charged surface of aggregates.

With this aim, this work compares the emulsification ability of three commercial lignins, by-products of the pulp and paper (kraft pulping) and bioethanol industries, which have been chemically modified via the Mannich reaction to be used as cationic emulsifiers. Initially, a comprehensive characterization of the different lignins has been carried out, and the emulsification ability of the resulting cationic surfactants has been assessed on prototype silicone oil-in-water emulsions in terms of droplet size distribution and viscous behavior. Afterwards, cationic bitumen emulsions stabilized by a modified lignin have been designed attending to the surfactant content and reagent molar ratio used for the amination reaction.

## 2. Materials and Methods

### 2.1. Raw Materials

A kraft lignin (referred to as KFT) from Sigma Aldrich (Madrid, Spain) and two lignin samples derived from the bioethanol process with two purification grades, provided by DONG Energy (Fredericia, Denmark) (referred to as BIOeth1 and BIOeth2, respectively), were used as precursors to obtain cationic lignins (C-KFT, C-BIOeth1 and C-BIOeth2). Bioethanol lignins are expected to be the solid residue fraction derived from wheat straw subjected to a hydrothermal pre-treatment followed by enzymatic hydrolysis and sugars fermentation to obtain bioethanol [24].

Lignin amination was performed following the Mannich reaction procedure, using tetraethylene pentamine (TEPA) and formaldehyde 37% as reagents, which were both supplied by Sigma Aldrich (Spain).

In the production of the model emulsions, silicone oil (FS100 from Esquim SA, Barcelona, Spain) with an approximate viscosity of 100 mPa·s at ambient temperature was chosen. This viscosity allowed us to easily reproduce at ambient conditions bitumen viscosity prior to its emulsification at high temperature, which is recommended to be less than 200 mPa·s before entering the colloid mill [20]. Subsequently, for the second part of the study, a bitumen with a penetration grade of 160–220 (Repsol SA, Madrid, Spain) was emulsified.

### 2.2. Preparations

Amination reaction was carried out under alkaline conditions at 60 °C for 4 h, previously dissolving lignin in distilled water at two pH values, 10 and 13. Formaldehyde (Fd) and tetraethylene pentamine (TEPA) were used as reagents for the amination of lignins (Lig) at three different Lig/TEPA/Fd stoichiometric molar ratios, 1/7/7, 1/14/14, and 1/28/28. Once reaction finished, the emulsification pH was reduced to 1 with hydrochloric acid (HCl) to activate new amine functional groups of modified lignins. Lignin amination by TEPA has been confirmed by FTIR analysis (as described below), suggesting that modification takes place with the inclusion of amine functional groups in the aromatic units [18].

The emulsification process of model silicone emulsions was performed with the IKA T25/25F homogenizer (Germany) at 20,000 rpm rotational speed for 4 min, and bitumen emulsification was carried out with an IKAT50/G45M at 10,000 rpm for 3 min. Silicone oil was emulsified at room temperature, whereas bitumen was blended with the aqueous phase (water and surfactant) to prepare a premixture at 90 °C prior to the emulsification stage. Two oily disperse phase (O) concentrations were studied, 60% and 70%, and cationic emulsifier (Surf) concentrations ranged from 0.25 to 0.75% Surf.

### 2.3. Tests and Measurements

Lignin composition was determined by standard analytical methods (National Renewable Energy Laboratory NREL/TP-510-42618). In order to determine the carbohydrate composition, samples were subjected to quantitative acidic hydrolysis in two steps. The obtained hydrolyzed liquids were then analyzed for sugar contents using an Agilent Technologies 1260 HPLC fitted to a refractive index detector and an Agilent Hi-PlexPb column operated at 70 °C with Milli-Q water as the mobile phase, which was pumped at a rate of 0.6 mL/min. Soluble lignin was also measured in the hydrolyzed liquids using a Lambda 365 UV/Vis spectrometer (Perkin Elmer, Boston, MA, USA) at 205 nm. The resulting solid residues obtained after the acid hydrolysis were considered acid-insoluble lignin (Klason lignin). Ash content was determined following the standard UNE 57050:2003. Finally, elemental analysis was conducted in an Elemental Thermo Flash EA1112 analyzer (Thermo Fisher Sci., MA, USA) following the standards EN 15104, EN 15296 and EN 15289.

The total phenols content of lignins was analyzed according to Jiménez-López et al. [25]. For that, lignins were previously dissolved in dimethylsulfoxide (DMSO). Then, the absorbance of a mixture with $Na_2CO_3$, Folin–Ciocalteau reagent, and lignin solutions was measured at 760 nm using a UV–Vis spectrophotometer (Lambda 365, PerkinElmer, Boston,

MA, USA). The total phenols content of samples was quantified using a calibration curve prepared from a standard solution of gallic acid (1–20 mg/L) and expressed as mg gallic acid equivalent (GAE)/g of lignin (on a dry basis).

Size exclusion chromatography analysis (SEC) was used to evaluate the average molecular weight (Mw), number average (Mn) and polydispersity (Mw/Mn) of the lignins. Lignin samples were examined by high-performance liquid chromatography (HPLC) using a column PLgel 10 µm MIXED B 300 × 7.5 mm operated at 70 °C. A column PLgel 10 µm guard 50 × 7.5 mm was also employed. N,N-dimethylformamide (DMF) pumped at a rate of 1 mL/min was employed as the mobile-phase. Polystyrene was used as the standard, with peak average molecular weights of 570, 8900, 62,500, 554,000 (Sigma-Aldrich, Spain). For detection, a G1362A refractive index (RI) detector (Agilent, Waldbronn, Germany) was used. Solubilization was almost complete; only approximately 5% of the samples remained undissolved, which it is considered a non-significant percentage of the samples for the determination of their molecular weight.

Fourier-transform infrared spectroscopy (FTIR) measurements were conducted for identifying the functional groups present in lignin samples. A JASCO FT/IR 460 Plus spectrometer (Jasco, Japan) was used to acquire the spectra from 4000 to 600 $cm^{-1}$, after 400 scans at 1 $cm^{-1}$ resolution.

Emulsion characterization was carried out by means of droplet size distribution measurements and viscosity curves at 30 °C. Particle size distribution tests were conducted in a Mastersizer 2000 Malvern (UK), at ambient temperature, according to the laser diffraction method. Steady flow viscous tests between 0.5 and 500 $s^{-1}$ were performed using a profiled-surface coaxial cylinder ($d_i$ = 26.657 mm and $d_o$ = 28.922 mm), to avoid wall-slip phenomena [26], in a controlled stress rheometer Anton Paar MCR301 (Graz, Austria).

### 3. Results and Discussion

#### 3.1. Lignin Characterization

Table 1 shows compositions obtained for the different raw lignin-rich products. As may be seen, the kraft lignin sample (KFT) shows a higher lignin concentration than the bioethanol-derived products (BIOeth1 and BIOeth2). This result is related to the higher concentrations of carbohydrate residues (around 23–24%) and ashes (9–10%) found for BIOeth1 and BIOeth2 samples. As known, the solid residual fraction after bioethanol production is enriched in lignin but with a certain contamination of carbohydrates derived from the non-hydrolyzed cellulose during the saccharification [8], among which glucose is the main sugar (Table 1).

**Table 1.** Composition of the different raw lignin-rich products.

| Sample | Klason Lignin (wt.%) | Soluble Lignin (wt.%) | Total Lignin (wt.%) | Glucose (wt.%) | Xylose (wt.%) | Arabinose (wt.%) | Ash (wt.%) | Elemental Analysis (wt.%) | | | | |
|--------|------|------|------|------|------|------|------|------|------|------|------|------|
| | | | | | | | | C | H | N | O | S |
| KFT | 89.08 | 5.14 | 94.22 | 0.92 | 1.61 | 0.20 | 2.4 | 61.2 | 6.72 | 2.18 | 26.9 | 1.8 |
| BIOeth1 | 60.59 | 3.35 | 63.94 | 23.61 | 5.68 | 0.26 | 10.4 | 49.1 | 5.16 | 2.37 | 33.4 | 0.13 |
| BIOeth2 | 62.12 | 3.62 | 65.74 | 24.06 | 5.49 | 0.17 | 9.4 | 49.0 | 5.21 | 2.75 | 34.0 | 0.12 |

During bioethanol production, hydrothermal pre-treatments solubilize a great part of hemicellulose fraction, resulting in pre-treated materials enriched mostly in cellulose and lignin [27]. This residual lignin in the pre-treated materials may unspecifically bind hydrolytic enzymes during the subsequent enzymatic hydrolysis [28], decreasing the saccharification yields and, consequently, increasing the content of carbohydrates non-hydrolyzed (mainly glucose) in bioethanol-derived lignin samples. Regarding the kraft lignin sample, the small amount of carbohydrates found (mainly xylose) is normally attributed to the lignin carbohydrates complexes and non-bounded sugars, as suggested by Alekhina et al. [29] and dos Santos et al. [30]. These authors have also described that the fast degradation of carbohydrates (mainly hemicelluloses as xylan) during kraft pulping, through the well-known peeling reactions, together with the low solubility of hemicelluloses in acid media during lignin precipitation from black liquors, are the factors that explain why xylose is the most abundant sugar in KFT (Table 1).

As for the higher ash content found for bioethanol-derived lignin samples compared with KFT (Table 1), the presence of ashes in kraft lignin is probably related to the inorganic compounds used in kraft pulping (e.g., NaOH and $Na_2S$) and the high amount of $Na_2SO_4$ salts formed during the acid precipitation step. However, the low content of ashes found for this lignin would indicate it has been subjected to a purification post-treatment. Conversely, the higher ash concentration observed for BIOeth1 and BIOeth2 samples cannot be explained by the isolation process of these lignins, where neither kraft cooking nor acid precipitation processes have been used. The reason for the higher ash concentration found in bioethanol lignins could be related to the content of inorganic compounds present in the raw material (e.g., wheat straw). It is known that the ash and silicate contents of different non-woody materials, such as wheat straw, are higher than in wood lignin [31]. In any case, the ash content would depend on post-treatments performed on the lignin-rich product, such as salt removal with a wash step using deionized water until pH 5 [29].

Another relevant difference observed among samples was the lower sulfur content measured in bioethanol samples, which is 10 times lower than that found in KFT (Table 1). This result is due to the use of $Na_2S$ during the kraft process, in which lignocellulosic raw material is treated with NaOH and $Na_2S$ in an aqueous solution at high temperature [32,33]. As a result, some ionizable groups as phenolic hydroxyls present in kraft lignin may become sulfonated during kraft pulping.

Regarding acid soluble lignin contents (Table 1), BIOeth1 and BIOeth2 lignin samples presented a lower amount than kraft lignin. This fact could be attributed to the hydrothermal pre-treatment used in bioethanol production that could have removed low molecular weight degradation products and hydrophilic derivatives of lignin, which are both components of the acid-soluble lignin [27]. On the other hand, the severity used during the kraft pulping along with the lignin precipitation step at low pH would favor the increase in the acid-soluble lignin content found in KFT [34].

FTIR spectroscopy was used to characterize the main functional components of lignin samples (Figure 1), showing all spectra the typical lignin bands according to previous studies [8,9,29,35]. At $3330$ $cm^{-1}$, a wide band attributed to O−H stretching vibration (both alcohol and phenolic hydroxyl groups) was displayed in all lignin spectra. The kraft lignin spectrum showed a broader band, moved toward $3250$ $cm^{-1}$, with higher intensity compared with bioethanol-derived lignin spectra, demonstrating a higher content of hydroxyl groups, mainly phenolic, possibly because of the major cleavage of alkyl–aryl ether linkages (β-O-4′ substructures) during the kraft pulping process [29,36]. Thus, the phenol content measured on samples agrees with this result, having KFT, BIOeth1 and BIOeth2, respectively, 366, 102 and 104 mg GAE/lignin. The bands at $2928$ $cm^{-1}$ and $2852$ $cm^{-1}$ correspond to the C−H stretching vibration in the $−CH_3$ and $−CH_2−$ groups, respectively, whereas the band at $1455$ $cm^{-1}$ is associated to the C−H asymmetric vibrations and deformation. The intensity of these bands was lower for kraft lignin compared with bioethanol-derived lignins, which could be related to a lower aliphatic group content, which is probably due to a higher aliphatic chain shortening during kraft pulping [36]. The higher content of hydroxyl groups, together with the aliphatic chain shortening observed in the kraft lignin spectrum, suggests a lignin with a major fragmented structure compared with bioethanol-derived lignins. Similar effects were described by Santos et al. [8] when bioethanol lignins from olive tree pruning were compared with alkaline lignins, either soda-anthraquinone or kraft lignins, the latter being more degraded.

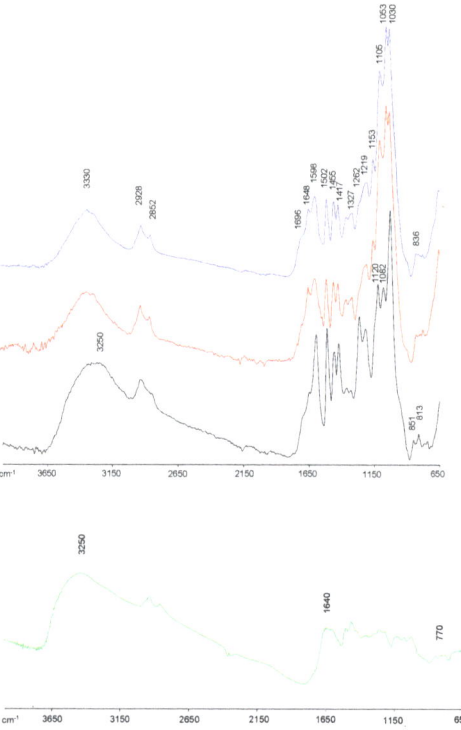

**Figure 1.** FTIR spectra, 4000–650 cm$^{-1}$ region, of KFT (black), BIOeth1 (red) and BIOeth2 (blue) lignin samples, and KFT lignin sample after amination reaction (green).

All lignin spectra presented the typical bands at 1598, 1502, and 1417 cm$^{-1}$ attributed to vibrations of lignin aromatic skeleton. Furthermore, the three spectra showed a shoulder at 1648 cm$^{-1}$ (conjugated C=O groups), especially visible in bioethanol lignins, which could be explained by lignin oxidation. Nevertheless, in the case of bioethanol lignins, these carbonyl groups may also be associated to the C=O stretching of amide bonds from cellulolytic enzymes and/or biological contaminations from the fermentative microorganisms [37–39]. This last hypothesis is supported by the higher nitrogen content found in bioethanol lignins samples (Table 1). Another shoulder at 1698 cm$^{-1}$, associated to unconjugated C=O groups stretching from lignin oxidation, was also observed in all spectra, although it can also be attributed to carbonyl groups in hemicelluloses impurities [29].

Regarding the presence of S, G or H units, the kraft lignin spectrum showed the typical pattern of softwood lignins, with pronounced bands at 1262 cm$^{-1}$ (G ring breathing with C=O stretching), 1219 cm$^{-1}$ (G ring breathing with C–C, C–O, and C=O stretching), 1030 cm$^{-1}$ (C–H bond deformation in G units) and the bands at 853 cm$^{-1}$ and 817 cm$^{-1}$ (C–H out of plane in position 2, 5 and 6 of G units) [29]. On the other hand, bioethanol-derived lignins showed the corresponding G units bands, together with S and H units bands at 1327 cm$^{-1}$ (S aromatic ring breathing) and at 836 cm$^{-1}$ (C–H out of plane in position 2 and 6 of S units and all positions of H units), which are typical of hardwood lignins as well as non-woody lignins (e.g., wheat straw) [8,9,35].

In agreement with the higher carbohydrate content (mainly glucose) determined in bioethanol lignins (Table 1), their spectra displayed cellulose bands at 1153, 1105, 1053, and 1030 cm$^{-1}$ overlapping some lignin bands. On the other hand, the kraft lignin spectrum showed a new band at 1082 cm$^{-1}$, which is associated to hemicelluloses. Furthermore, as previously commented, the band intensity at 1698 cm$^{-1}$ could also be partly correlated to hemicelluloses.

After amination reaction, the structural units appear different as observed by the following evidence on the modified Kraft lignin spectrum: (a) a broader and stronger band at 3250 cm$^{-1}$, in which N-H stretching vibrations could be contributing, as well as the presence of a newly developed peak centered at 770 cm$^{-1}$ associated to N-H wagging out of plane; (b) a broader band appeared at 1640 cm$^{-1}$, arising from the N-H bending vibrations in a single bond NH$_2$ structure; and (c) the C–H vibrations from the aromatic skeleton of lignin showed a lesser definition, such as 1598, 1502, 1455 and 1417 cm$^{-1}$ [18,40,41]. This would suggest that modification, with the inclusion of an amine functional group, has occurred in their aromatic unit.

Finally, Figure 2 and Table 2 show, respectively, lignin molecular weight distributions and their average molecular weight (Mw), number average (Mn), and polydispersity (Mw/Mn). As seen in Table 2, all lignins showed low molecular weights with very similar polydispersity values. Nevertheless, the kraft lignin sample showed a lower molecular weight than lignin samples from bioethanol production (BIOeth1 and BIOeth2). The lower molecular weight of kraft lignin corroborates with the higher phenolic content observed by the FTIR spectroscopy of this lignin, compared with bioethanol lignins, which supports a major lignin depolymerization during the kraft pulping process. In this sense, previous studies indicated that the molecular weight of kraft lignin could be greatly reduced due to an extensive cleavage of alkyl-aryl ether linkages of β-O-4′ bonds during kraft pulping [29,42]. On the contrary, lignins from bioethanol production are usually found to be less degraded [8].

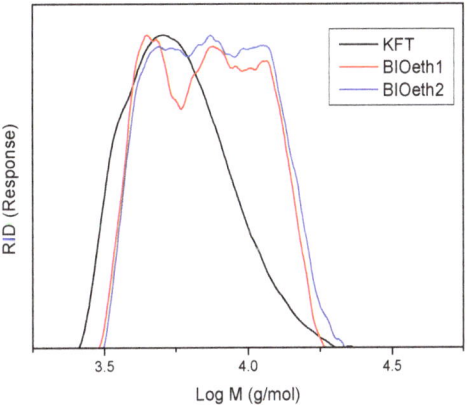

**Figure 2.** Molecular weight distribution of KFT (black), BIOeth1 (red) and BIOeth2 (blue) lignin samples.

**Table 2.** Average molecular weight (Mw), number average (Mn) and polydispersity (Mw/Mn) of the lignin samples.

| Sample | Mn (g/mol) | Mw (g/mol) | Polydispersity |
|---|---|---|---|
| KFT | 5366 | 6248 | 1.164 |
| BIOeth1 | 6618 | 7869 | 1.189 |
| BIOeth2 | 6878 | 8217 | 1.195 |

### 3.2. Lignin as Cationic Surfactants of Model Emulsions

Emulsion properties such as droplet size distribution (DSD) and rheology are closely linked to emulsifier performance. Optimization of the emulsifier was initially carried out on model silicone–oil-based emulsions by studying the effect of the lignin source and reaction pH for a selected reagent ratio of 1/14/14 (Lig/TEPA/Fd reagents), which was calculated according to the previous lignin characterization. Figure 3 shows the effect of lignin source on the emulsification ability of the cationic surfactants prepared under alkaline conditions (pH = 10 and 13). When reaction at pH of 13 was used for lignin amination, no matter the selected lignin source, a 0.5% cationic lignin was able to stabilize the 60% and 70% oil emulsions formulated at pH = 1. Systems emulsified by bioethanol-derived cationic surfactants (C-BIOeth1 and C-BIOeth2) showed bimodal distributions, obtaining a narrower droplet size distribution (DSD) for the emulsions stabilized by C-BIOeth2 (i.e., the most purified lignin by-product from the bioethanol process). Conversely, emulsion stabilized by cationic kraft lignin (C-KFT) showed a monomodal DSD with a shoulder at low particle size.

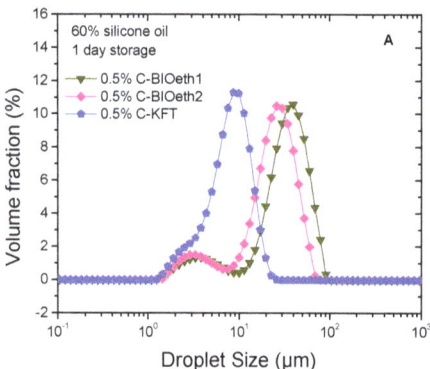 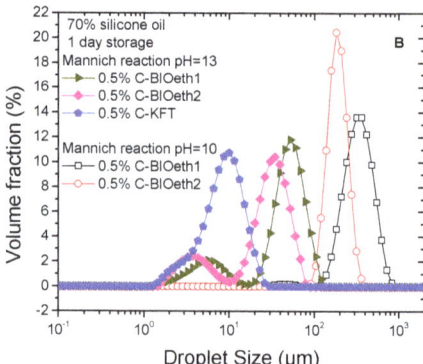

**Figure 3.** Droplet size distribution of emulsions formulated with 60% (**A**) and 70% silicone oil (**B**) and stabilized by 0.5% surfactant. Emulsification ability of cationic surfactants as a function of their lignin source and pH of amination.

The mean particle size of emulsions was determined as Sauter ($D_{3,2}$) and De-Brouckere or volumetric ($D_{4,3}$) diameters [43]:

$$D_{3,2} = \frac{\sum_i n_i d_i^3}{\sum_i n_i d_i^2},$$
(1)

$$D_{4,3} = \frac{\sum_i n_i d_i^4}{\sum_i n_i d_i^3}$$
(2)

where $n_i$ is the number of droplets with diameter $d_i$. Table 3 shows that C-KFT leads to emulsions with much lower Sauter and volumetric mean diameters. Furthermore, an increase in oil concentration from 60% to 70% hardly modifies the values of $D_{3,2}$ and $D_{4,3}$ no matter the lignin considered. Likewise, the emulsion particle size remained almost constant after one week of storage time at room temperature (Table 3). Conversely, a less alkaline medium (pH = 10) for the amination reaction significantly reduced lignin's emulsification ability, shifting DSD curves toward higher droplet sizes (Figure 3B and Table 3).

The viscous behavior of prepared emulsions is shown in Figure 4. Systems presented a shear thinning non-Newtonian behavior characterized by a power-law decrease in viscosity with shear rate, which is more evident for the most oil-concentrated (and, therefore, more viscous) emulsions (Figure 4B). This behavior is well known in flocculated emulsions, in which the slope of the shear-thinning region is higher as the flocculation degree increases (Santos et al., 2015). Thus, Figure 4B shows, for cationic lignins synthetized at pH13, that emulsions stabilized by bioethanol-derived lignins present viscosity curves steeper (i.e., more dependent on shear rate) than C-KFT-stabilized systems. However, a decrease in the slopes of the shear-thinning region is found with storage time, suggesting a weakening of flocculation degree and a slow destabilization process of the emulsion that, instead, does not lead to an eventual droplet coalescence, as may be deduced from Table 3.

**Table 3.** Mean droplet diameter of cationic silicone emulsions prepared at pH = 1 and with a 0.5% surfactant (reagent molar ratio 1/14/14).

| Lignin Type | Reaction pH | Oil Conc. (wt.%) | $D_{3,2}$ (µm) Storage Time | | $D_{4,3}$ (µm) Storage Time | |
|---|---|---|---|---|---|---|
| | | | 1 Day | 1 Week | 1 Day | 1 Week |
| BIOeth1 | 13 | 60 | 16.9 | - | 37.3 | - |
| BIOeth2 | 13 | 60 | 14.2 | - | 27.4 | - |
| KFT | 13 | 60 | 6.8 | - | 9.2 | - |
| BIOeth1 | 10 | 70 | 311.9 | - | 376.5 | - |
| BIOeth2 | 10 | 70 | 194.3 | - | 207.5 | - |
| BIOeth1 | 13 | 70 | 18.7 | 17.3 | 49.7 | 49.3 |
| BIOeth2 | 13 | 70 | 12.8 | 12.9 | 31.0 | 32.9 |
| KFT | 13 | 70 | 7.3 | 7.8 | 10.0 | 10.5 |

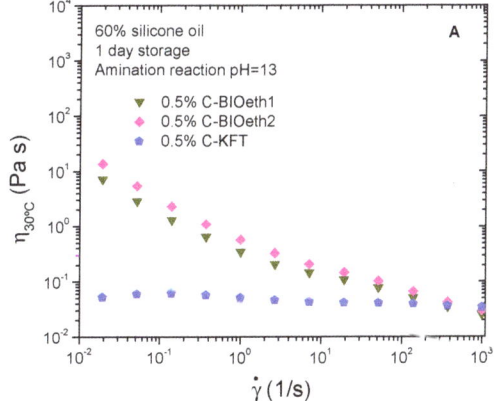 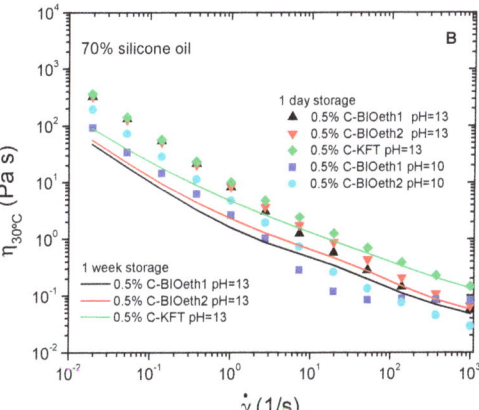

**Figure 4.** Viscous behavior of silicone emulsions stabilized by cationic lignin. Effect of the lignin source of cationic surfactants synthetized, pH of amination reaction and storage time on emulsions containing 60% (**A**) and 70% oil (**B**).

Interestingly, when the oil concentration is 60%, which is far below the expected maximum packing fraction for the emulsion disperse phase, cationic kraft lignin leads to weakly flocculated emulsions as may be deduced from the almost Newtonian behavior exhibited by this system, unlike bioethanol-derived cationic lignins (Figure 4A). Flocculation is known to increase the apparent dispersed phase volume, along with the formation of non-spherical aggregates. Both factors may contribute to the development of a non-Newtonian viscous response [44]. Under such conditions, with a less packed disperse phase in the emulsion (i.e., with a higher distance among droplets), only C-BIOeth1 and C-BIOeth2 would be able to build up an extended three-dimensional network formed by droplets interconnected by the lignin surfactant. This network seems to be more developed for the C-BIOeth2 stabilized emulsion as may be deduced from its steeper viscous flow curve (Figure 3A).

Results obtained would be in good agreement with the structures and compositions of the above-described lignin-rich products. Thus, cationic surfactants obtained from bioethanol process are expected to have a higher molecular weight (Table 2) than kraft lignin cationic-derived surfactant. Likewise, the contamination of these lignins by sugars may have a strong influence on the emulsion rheology, since polysaccharides are known to have a thickening effect with little surface activity, increasing the viscosity of the continuous aqueous phase of the emulsions [44,45]. Together, the higher lignin molecular weight and the thickening effect of carbohydrate contamination are expected to favor droplet flocculation.

Conversely, the lower molecular weight of kraft lignin (with a more fragmented structure compared with bioethanol-derived lignins) would lead to less flocculated systems with a lower droplet size, suggesting a higher interfacial activity for C-KFT. This fact, together with its higher purification degree, with about 94% lignin concentration (Table 1), makes cationic kraft lignin the best candidate to stabilize bitumen emulsions.

### 3.3. Formulation of Lignin-Based Cationic Bituminous Emulsions

According to the previous results, cationic kraft lignin showed the best performance as an emulsifier of silicone oil model emulsions, being selected as a potential cationic surfactant for bitumen emulsions. Regarding its surfactant activity, previously, it was found that the surface tension of this cationic lignin decreases with the increase in concentration up to 0.625% surfactant, and both polar and dispersive components of surface energy remain constant for higher concentrations [19]. As a result, the effect of surfactant concentration was initially assessed for the design of such emulsions in the range of 0.25–0.75%. Figure 5A shows that C-KFT successfully stabilizes water-in-oil emulsion containing 60% bitumen (O) at surfactant concentrations between 0.25 and 0.75%. In all cases, DSD curves show a trend to a bimodal distribution with peaks located at about 15 and 95 μm. Likewise, even though mean droplet diameters are similar for the three surfactant concentrations (Table 4), the smallest values of $D_{3,2}$ and $D_{4,3}$ are found for the emulsion stabilized with 0.75% C-KFT. All emulsions remained visually stable for at least one week, which is the shortest storage time required for this type of emulsions to be used in road applications.

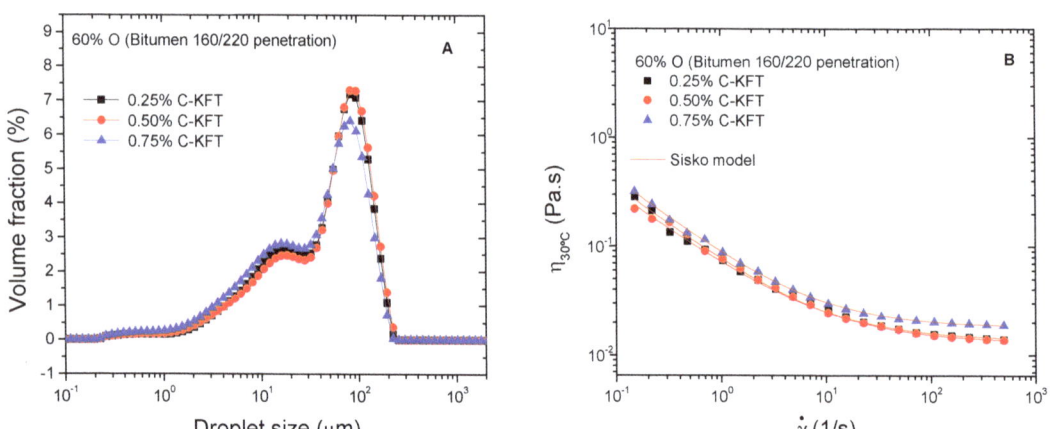

**Figure 5.** Effect of C-KFT concentration on droplet size distribution (**A**) and viscosity (**B**) of 60% bitumen emulsions.

**Table 4.** Mean droplet diameter and Sisko model parameters of 60% bitumen emulsions as a function of C-KFT concentration (reagent molar ratio 1/14/14).

| Surfactant Conc. (wt.%) | $D_{3,2}$ (μm) | $D_{4,3}$ (μm) | $\eta_\infty$ (Pa s) | $k$ (Pa s$^n$) | $n$ (-) |
|---|---|---|---|---|---|
| 0.25 | 11.7 | 60.9 | 0.014 | 0.06 | 0.29 |
| 0.50 | 11.4 | 63.7 | 0.013 | 0.06 | 0.27 |
| 0.75 | 9.1 | 53.9 | 0.018 | 0.07 | 0.23 |

Similarly, emulsion viscosities hardly change with surfactant concentration (Figure 5B). All flow curves display a non-Newtonian behavior with a shear thinning region at low shear rates followed by a trend to reach a high-shear-rate limiting viscosity, which corresponds to the shear-induced deflocculation process. The observed viscous behavior may be described by the Sisko model:

$$\eta = \eta_\infty + k\dot{\gamma}^{n-1}, \tag{3}$$

where $k$ and $n$ are, respectively, the consistency and flow indexes, and $\eta_\infty$ is the high-shear-rate limiting viscosity. As may be seen in Table 4, the Sisko model parameters slightly change with surfactant concentration, showing a trend to decrease flow index values with increasing concentration. Compared with previous model emulsions containing the same oil concentration, bituminous emulsions present higher viscosity and a more developed non-Newtonian character, which is characteristic of a more complex microstructure.

Finally, with the aim of optimizing surfactant properties, the effect of the reagent molar ratios selected for lignin aminations was assessed with two additional ratios, below and above the previously studied reagent ratio 1/14/14 (MKL/Am/Fd). As may be seen in Figure 6A, an increase in reagent ratios from 1/7/7 to 1/28/28 does not lead to remarkable changes in DSD curves, although the number of small droplets tends to increase above 1/7/7, with the development of the peak located around 15 μm. Interestingly, the smallest mean droplet diameters were obtained for a reagent ratio of 1/14/14 (Table 5).

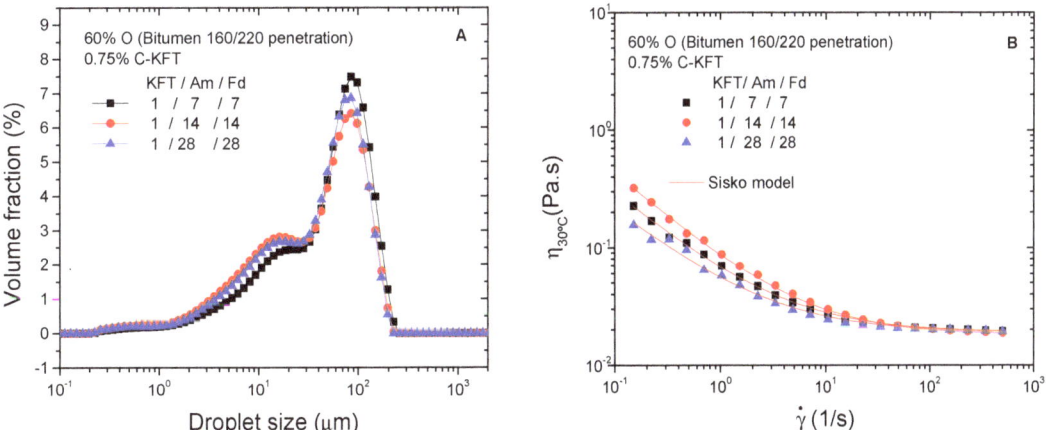

**Figure 6.** Effect of reagent molar ratio on droplet size distribution (**A**) and viscosity (**B**) of emulsions containing 60% bitumen and 0.75% C-KFT.

**Table 5.** Mean droplet diameters and Sisko model parameter of 60% bitumen emulsions as a function of C-KFT reagent molar ratio.

| Reagent Molar Ratio (KFT/TEPA/Fd) | $D_{3,2}$ (μm) | $D_{4,3}$ (μm) | $\eta_\infty$ (Pa s) | $k$ (Pa s$^n$) | $n$ (-) |
|---|---|---|---|---|---|
| 1/7/7 | 12.20 | 63.84 | 0.019 | 0.05 | 0.25 |
| 1/14/14 | 9.12 | 53.85 | 0.018 | 0.07 | 0.23 |
| 1/28/28 | 10.24 | 55.10 | 0.019 | 0.04 | 0.28 |

A shear-thinning behavior followed by a constant high shear-limiting viscosity was found for all samples (Figure 6B). Comparing the values of Sisko model parameters (Table 5), high-shear-rate limiting viscosities are hardly affected by reagent stoichiometric. The highest consistency index (k) and slope for the shear-thinning region (i.e., the lowest value of n) were found for the surfactant synthetized with a molar ratio of 1/14/14, whilst the lowest ones corresponded to the system stabilized with a surfactant prepared with a 1/28/28 reagent ratio. In any case, all reactions, with stoichiometric values between 1/7/7 and 1/28/28, were considered successful as deduced from emulsifier solubility under the acidic environment and from the already demonstrated performance as emulsifiers.

## 4. Conclusions

The kraft lignin used in this work as a source of cationic surfactants is characterized by a high purity: about 94% lignin concentration. Conversely, less purified bioethanol-derived samples, with a lignin content around 65%, also contain high concentrations of carbohydrate residues (23–24%) and ashes (9–10%). FTIR spectroscopy conducted on kraft lignin, compared with bioethanol-derived products, supports a major lignin depolymerization during the kraft pulping process, showing a lower content of aliphatic groups and higher content of hydroxyl groups. As a result, kraft lignin presents a more fragmented structure with a lower molecular weight.

On these grounds, the cationic surfactants obtained from the bioethanol process are expected to have a higher molecular weight than the kraft lignin cationic surfactant, which, together with the thickening effect of carbohydrate contamination, would favor droplet flocculation. The resultant emulsions would be stabilized by a combination of surfactant interfacial (electrostatic) interactions, due to the positive charge of the protonated amine groups, and steric interactions related to the high molecular weight of lignin and the presence of carbohydrates located in the emulsion continuous phase. A decrease in the alkalinity used for the amination reaction of these lignins, from pH = 13 to 10, did not improve their performance as cationic emulsifiers, leading to dispersions with large droplet sizes.

Conversely, the lower molecular weight and higher purification degree of raw kraft lignin has led to less flocculated emulsions with a lower droplet size, suggesting a higher interfacial activity for C-KFT. As a result, cationic kraft lignin has been selected as the best candidate to stabilize bitumen emulsions.

C-KFT has successfully stabilized oil-in-water emulsions containing 60% bitumen, with surfactant concentrations between 0.25 and 0.75%. In all cases, DSD curves have shown a trend to a bimodal distribution with peaks located at about 15 and 95 µm, and emulsions exhibited a non-Newtonian flow behavior, with a shear-thinning region followed by a trend to reach a constant viscosity at a high shear rate. Amination reactions, performed at pH = 13 and stoichiometries between 1/7/7 and 1/28/28, were considered successful as deduced from the solubility of the modified (cationic) emulsifier under an acidic environment (aqueous phase of the emulsion) and their demonstrated performance as cationic emulsifiers. In this sense, a more depolymerized lignin (i.e., with lower molecular weight) would result in a surfactant that has a higher positive charge per gram added to the emulsion. That weight of emulsifier would be composed by more molecules of lignin with protonated amine groups.

**Author Contributions:** A.Y.: investigation, methodology. P.P.: conceptualization, writing—original draft, project administration, funding acquisition. F.J.N.: writing—review and editing. R.M.-S.: investigation, methodology. D.I.: methodology, writing—review and editing, funding acquisition. M.E.E.: writing—review and editing. Project administration, funding acquisition. All authors have read and agreed to the published version of the manuscript.

**Funding:** This research is part of SUP&R ITN (FP7-PEOPLE-2013-ITN, Grant agreement 607524) and GreenAsphalt project (ref. 802C1800001) co-funded 80% by FEDER European Programme and 20% by Junta de Andalucía (Consejería de Economía, Conocimiento, Empresas y Unversidades/Agencia-IDEA). The authors also wish to thank Comunidad de Madrid and MCIU/AEI/FEDER, UE via grant number SUSTEC-CM S2018/EMT-4348 and RTI2018-096080-B-C22, respectively. The contribution of COST Action LignoCOST (CA17128), supported by COST (European Cooperation in Science and Technology), in promoting the interaction, exchange of knowledge and collaborations in the field of lignin valorization is gratefully acknowledged.

*Polymers* **2022**, *14*, 2879

**Institutional Review Board Statement:** Not applicable.

**Informed Consent Statement:** Not applicable.

**Data Availability Statement:** The data presented in this study are available on request from the corresponding author.

**Conflicts of Interest:** The authors declare no conflict of interest.

## References

1.  Bolker, I.H. *Natural and Synthetic Polymers: An Introduction*; Marcel Dekker, Inc.: New York, NY, USA, 1974. [CrossRef]
2.  Martínez, A.T.; Ruíz-Dueñas, J.; Martínez, M.J.; Del Río, J.C.; Gutiérrez, A. Enzymatic delignification of plant cell wall: From nature to mill. *Curr. Opin. Biotechnol.* **2009**, *20*, 348–357. [CrossRef] [PubMed]
3.  Ralph, J.; Lundquist, K.; Brunow, G.; Lu, F.; Kim, H.; Schatz, P.F.; Marita, J.M.; Hadfield, R.D.; Ralph, S.A.; Christensen, J.H. Lignins: Natural polymers from oxidative coupling of 4-hydroxyphenyl-propanoids. *Phytochem. Rev.* **2004**, *3*, 29–60. [CrossRef]
4.  Tejado, A.; Pena, C.; Labidi, J.; Echeverria, J.; Mondragon, I. Physico-chemical characterization of lignins from different sources for use in phenol–formaldehyde resin synthesis. *Bioresour. Technol.* **2007**, *98*, 1655–1663. [CrossRef] [PubMed]
5.  Laurichesse, S.; Avérous, L. Chemical modification of lignins: Towards biobased polymers. *Prog. Polym. Sci.* **2014**, *39*, 1266–1290. [CrossRef]
6.  Mikkonen, K.S. Strategies for structuring diverse emulsion systems by using wood lignocellulose-derived stabilizers. *Green Chem.* **2020**, *22*, 1019–1037. [CrossRef]
7.  Lourenço, A.; Pereira, H. *Compositional Variability of Lignin in Biomass, Lignin—Trends and Applications, Matheus Poletto*; IntechOpen: London, UK, 2017. [CrossRef]
8.  Santos, J.I.; Fillat, Ú.; Martín-Sampedro, R.; Eugenio, M.E.; Negro, M.J.; Ballesteros, I.; Rodríguez, A.; Ibarra, D. Evaluation of lignins from side-streams generated in an olive tree pruning-based biorefinery: Bioethanol production and alkaline pulping. *Int. J. Biol. Macromol.* **2017**, *105*, 238e251. [CrossRef]
9.  Martín-Sampedro, R.; Santos, J.I.; Fillat, Ú.; Wicklein, B.; Eugenio, M.E.; Ibarra, D. Characterization of lignins from *Populus alba* L. generated as by-products in different transformation processes: Kraft pulping, organosolv and acid hydrolysis. *Int. J. Biol. Macro.* **2019**, *126*, 18–29. [CrossRef]
10. Carvajal, J.C.; Gómez, Á.; Cardona, C.A. Comparison of lignin extraction processes: Economic and environmental assessment. *Bioresour. Technol.* **2016**, *214*, 468–476. [CrossRef]
11. Bajwa, D.S.; Pourhashem, G.; Ullah, A.H.; Bajwa, S.G. A concise review of current lignin production, applications, products and their environmental impact. *Ind. Crop. Prod.* **2019**, *139*, 111526. [CrossRef]
12. Brännvall, E. Overview of pulp and paper processes. In *Pulping Chemistry and Technology*; Ek, M., Gellerstedt, G., Henriksson, G., Eds.; De Gruyter: Berlin, Germany; New York, NJ, USA, 2009. [CrossRef]
13. Ľudmila, H.; Michal, J.; Andrea, Š.; Aleš, H. Lignin, potential products and their market value. *Wood Res.* **2015**, *60*, 973–986.
14. Susmozas, A.; Martín-Sampedro, R.; Eugenio, M.E.; Iglesias, R.; Manzanares, P.; Moreno, A.D. Process strategies for the transition of 1G to advanced bioethanol production. *Processes* **2020**, *8*, 1310. [CrossRef]
15. Martín-Sampedro, R.; Santos, J.I.; Eugenio, M.E.; Wicklein, B.; Jiménez-López, L.; Ibarra, D. Chemical and thermal analysis of lignin streams from *Robinia pseudoacacia* L. generated during organosolv and acid hydrolysis pre-treatments and subsequent enzymatic hydrolysis. *Int. J. Biol. Macromol.* **2019**, *140*, 311–322. [CrossRef] [PubMed]
16. Du, X.; Li, J.; Lindström, M.E. Modification of industrial softwood kraft lignin using Mannich reaction with and without phenolation pretreatment. *Ind. Crop. Prod.* **2014**, *52*, 729–735. [CrossRef]
17. Lesueur, D. *Polymer Modified Bitumen Emulsions (PMBEs), Polymer Modified Bitumen*; Elsevier: Amsterdam, The Netherlands, 2011. [CrossRef]
18. Yuliestyan, A.; García-Morales, M.; Moreno, E.; Carrera, V.; Partal, P. Assessment of modified lignin cationic emulsifier for bitumen emulsions used in road paving. *Mater. Des.* **2017**, *131*, 242–251. [CrossRef]
19. Yuliestyan, A.; Gavet, T.; Marsac, P.; García-Morales, M.; Partal, P. Sustainable asphalt mixes manufactured with reclaimed asphalt and modified-lignin-stabilized bitumen emulsions. *Constr. Build. Mater.* **2018**, *173*, 662–671. [CrossRef]
20. Read, J.; Whiteoak, D. *The Shell Bitumen Handbook*, 5th ed.; Thomas Teldford Publishing: London, UK, 2003. [CrossRef]
21. Ronald, M.; Luis, F.P. Asphalt emulsions formulation: State-of-the-art and dependency of formulation on emulsions properties. *Constr. Build. Mater.* **2016**, *123*, 162–173. [CrossRef]
22. Mannich, C.; Krösche, W. Ueber ein Kondensationsprodukt aus Formaldehyd, Ammoniak und Antipyrin. *Arch. Pharm.* **1912**, *250*, 647–667. [CrossRef]
23. Li, J.J. Mannich reaction. In *Name Reactions*; Springer: Berlin/Heidelberg, Germany, 2009. [CrossRef]
24. Landa, P.A.; Gosselink, R.J.A. Lignin-Based Bio-Asphalt. Patent WO 2019/092278, 16 May 2019.
25. Jiménez-López, L.; Martín-Sampedro, R.; Eugenio, M.E.; Santos, J.I.; Sixto, H.; Cañellas, I.; Ibarra, D. Co-production of soluble sugars and lignin from short rotation white poplar and black locust crops. *Wood Sci. Technol.* **2020**, *54*, 1617–1643. [CrossRef]
26. Franco, J.M.; Gallegos, C.; Barnes, H.A. On Slip Effects in Steady-State Flow Measurements of Oil-in-Water Food Emulsions. *J. Food Eng.* **1998**, *36*, 89–102. [CrossRef]

27. Alvira, P.; Tomás-Pejó, E.; Ballesteros, M.; Negro, M.J. Pretreatment technologies for an efficient bioethanol production process based on enzymatic hydrolysis: A review. *Bioresour. Technol.* **2010**, *101*, 4851–4861. [CrossRef] [PubMed]
28. Berlin, A.; Balakshin, M.; Gilkes, N.; Kadla, J.; Maximenko, V.; Kubo, S. Inhibition of cellulase, xylanase and beta-glucosidase activities by softwood lignin preparations. *J. Biotechnol.* **2006**, *125*, 198–209. [CrossRef] [PubMed]
29. Alekhina, M.; Ershova, O.; Ebert, A.; Heikkinen, S.H. Sixta Softwood kraft lignin for value-added applications: Fractionation and structural characterization. *Ind. Crop. Prod.* **2015**, *66*, 220–228. [CrossRef]
30. dos Santos, P.S.B.; Erdocia, X.; Gatto, D.A.; Labidi, J. Characterisation of Kraft lignin separated by gradient acid precipitation. *Ind. Crop. Prod.* **2014**, *55*, 149–154. [CrossRef]
31. Gosselink, R.J.A.; Abächerli, A.; Semke, H.; Malherbe, R.; Käuper, P.; Nadif, A.; van Dama, J.E.G. Analytical protocols for characterisation of sulphur-free lignin. *Ind. Crop. Prod.* **2004**, *19*, 271–281. [CrossRef]
32. Belgacem, M.N.; Gandini, A. *Monomers, Polymers and Composites from Renewable Resources*; Elsevier: Amsterdam, The Netherlands, 2008. [CrossRef]
33. Collins, M.N.; Nechifor, M.; Tanasă, F.; Zănoagă, M.; McLoughlin, A.; Stróżyk, M.A.; Culebras, M.; Teacă, C.-A. Valorization of lignin in polymer and composite systems for advanced engineering applications—A review. *Int. J. Biol. Macromol.* **2019**, *131*, 828–849. [CrossRef]
34. Yasuda, S.; Fukushima, K.; Kakehi, A. Formation and chemical structures of acid-soluble lignin I: Sulfuric acid treatment time and acid-soluble lignin content of hardwood. *J. Wood. Sci.* **2001**, *47*, 69–72. [CrossRef]
35. del Río, J.C.; Gutiérrez, A.; Rodríguez, I.M.; Ibarra, D.; Martínez, A.T. Composition of non-woody plant lignins and cinnamic acids by Py-GC/MS, Py/TMAH and FT-IR. *J. Anal. Appl. Pyrolysis* **2007**, *79*, 39–46. [CrossRef]
36. Ibarra, D.; Chávez, M.I.; Rencoret, J.; del Río, J.C.; Gutiérrez, A.; Romero, J.; Camarero, S.; Martínez, M.J.; Jiménez-Barbero, J.; Martínez, A.T. Lignin modification during Eucalyptus globulus kraft pulping followed by totally chlorine free bleaching: A two-dimensional nuclear magnetic resonance, fourier transform infrared, and pyrolysis-gas chromatography/mass spectrometry study. *J. Agric. Food Chem.* **2007**, *55*, 3477–3490. [CrossRef]
37. Ibarra, D.; del Río, J.C.; Gutiérrez, A.; Rodríguez, I.M.; Romero, J.; Martínez, M.J.; Martínez, A.T. Isolation of high-purity residual lignins from eucalypt paper pulps by cellulase and proteinase treatments followed by solvent extraction. *Enzyme Microb. Technol.* **2004**, *35*, 173–181. [CrossRef]
38. Santos, J.I.; Martín-Sampedro, R.; Fillat, Ú.; Oliva, J.M.; Negro, M.J.; Ballesteros, M.; Eugenio, M.E.; Ibarra, D. Evaluating lignin-rich residues from biochemical ethanol production of wheat straw and olive tree pruning by FTIR and 2D-NMR. *Int. J. Pol. Sci.* **2015**, *2015*, 314891. [CrossRef]
39. Santos, J.I.; Fillat, Ú.; Martín-Sampedro, R.; Ballesteros, I.; Manzanares, P.; Ballesteros, M.; Eugenio, M.E.; Ibarra, D. Lignin-enriched fermentation residues from bioethanol production of fast growing poplar and forage sorghum. *Bioresources* **2015**, *10*, 5215–5232. [CrossRef]
40. Liu, Z.; Lu, X.; An, L.; Xu, C. A novel cationic lignin-amine emulsifier with high performance reinforced via phenolation and Mannich reactions. *BioResources* **2016**, *11*, 6438–6451. [CrossRef]
41. Jiao, G.J.; Peng, P.; Sun, S.L.; Geng, Z.C.; She, D. Amination of biorefinery technical lignin by Mannich reaction for preparing highly efficient nitrogen fertilizer. *Int. J. Biol. Macro.* **2019**, *177*, 544–554. [CrossRef] [PubMed]
42. Eugenio, M.E.; Martín-Sampedro, R.; Santos, J.I.; Wicklein, B.; Martín, J.A.; Ibarra, D. Properties versus application requirements of solubilized lignins from an elm clone during different pre-treatments. *Int. J. Biol. Macromol.* **2021**, *181*, 99–111. [CrossRef] [PubMed]
43. Piacentini, E. Droplet Size. In *Encyclopedia of Membranes*; Drioli, E., Giorno, L., Eds.; Springer: Berlin/Heidelberg, Germany, 2016. [CrossRef]
44. Santos, J.; Calero, N.; Guerrero, A.; Muñoz, J. Relationship of rheological and microstructural properties with physical stability of potato protein-based emulsions stabilized by guar gum. *Food Hydrocoll.* **2015**, *44*, 109–114. [CrossRef]
45. Erçelebi, E.A.; Ibanoğlu, E. Rheological properties of whey protein isolate stabilized emulsions with pectin and guar gum. *Eur. Food Res. Technol.* **2009**, *229*, 281–286. [CrossRef]

*polymers*

*Article*

# Changes in Crystal Structure and Accelerated Hydrolytic Degradation of Polylactic Acid in High Humidity

Yutaka Kobayashi [1,*] , Tsubasa Ueda [2], Akira Ishigami [1,2] and Hiroshi Ito [1,2,*]

1   Research Center for GREEN Materials and Advanced Processing (GMAP), 4-3-16 Jonan,
    Yonezawa 992-8510, Japan; akira.ishigami@yz.yamagata-u.ac.jp
2   Graduate School of Organic Materials Science, Yamagata University, 4-3-16 Jonan, Yonezawa 992-8510, Japan;
    t211686d@st.yamagata-u.ac.jp
*   Correspondence: kobayashi.y@yz.yamagata-u.ac.jp (Y.K.); ihiroshi@yz.yamagata-u.ac.jp (H.I.);
    Tel.: +81-238-26-3430 (Y.K.); +81-238-26-3081 (H.I.)

**Abstract:** Highly crystallized polylactic acid (PLA) is suitable for industrial applications due to its stiffness, heat resistance, and dimensional stability. However, crystal lamellae in PLA products might delay PLA decomposition in the environment. This study clarifies how the initial crystal structure influences the hydrolytic degradation of PLA under accelerated conditions. Crystallized PLA was prepared by annealing amorphous PLA at a specific temperature under reduced pressure. Specimens with varied crystal structure were kept at 70 °C and in a relative humidity (RH) of 95% for a specific time. Changes in crystal structure were analyzed using differential calorimetry and wide-angle X-lay diffraction. The molecular weight (MW) was measured with gel permeation chromatography. The crystallinity of the amorphous PLA became the same as that of the initially annealed PLA within one hour at 70 °C and 95% RH. The MW of the amorphous PLA decreased faster even though the crystallinity was similar during the accelerated degradation. The low MW chains of the amorphous PLA tended to decrease faster, although changes in the MW distribution suggested random scission of the molecular chains for initially crystallized PLA. The concentrations of chain ends and impurities, which catalyze hydrolysis, in the amorphous region were considered to be different in the initial crystallization. The crystallinity alone does not determine the speed of hydrolysis.

**Keywords:** biopolymers; hydrolytic degradation; crystal structure

**Citation:** Kobayashi, Y.; Ueda, T.; Ishigami, A.; Ito, H. Changes in Crystal Structure and Accelerated Hydrolytic Degradation of Polylactic Acid in High Humidity. *Polymers* 2021, *13*, 4324. https://doi.org/10.3390/polym13244324

Academic Editor: Piotr Bulak

Received: 28 October 2021
Accepted: 8 December 2021
Published: 10 December 2021

## 1. Introduction

Polylactic acid (PLA), as a bio-polymer, is expected to assist the transition to a carbon-neutral world. A low-molecular-weight (MW) PLA was prepared by ring-opening polymerization in 1932, and a high-MW PLA was made from purified raw materials in 1954 [1]. Moreover, in 1954, polypropylene (PP) was discovered [2]. The global production capacity of PLA in 2020 was 390,000 tons [3], while that of PP was 90 million tons [4]. One of the reasons for such a large difference is the performance of polymeric material.

The goal of PLA researchers has been to achieve long-term durability, molding processability, and thermal stability. Thus, the optical purity of the L-form has been increased, the polymerized product has been purified, and the chain ends of the polymer have been protected [5]. Currently, high-performance PLA with high crystallinity is manufactured. However, the improvement in durability indicates a decrease in degradability in the environment. Hence, PLA has caused plastics waste. In particular, PLA is not marine degradable [6].

After a long period in the ocean, PLA becomes a microplastic, similar to PP. The environmental impact of microplastics can be divided into the absorption of toxic chemicals and their own toxicity. Since PLA is less hydrophobic, it adsorbs fewer polynuclear aromatics [7], but the biological activity of the sediments made from microplastics may affect organisms [8]. Thus, there is a demand for a plastic with high marine decomposition properties. Not limited to synthetic or natural polymers, highly degradable plastics are characterized by a low glass transition temperature (Tg), low fusion heat, and low hydrophobicity [9]. It is difficult to achieve both degradability and mechanical performance.

Molded products of amorphous PLA will be deformed above the Tg of 65 °C. To stabilize the dimensions of products at high temperatures, sufficiently crystallized PLA is utilized. Factors that control the crystallinity of PLA are the processing conditions, the optical purity of polymer chains, and doped additives such as plasticizers and nucleating agents. Crystallized PLA shows that enzymatic degradation occurs at the surface of the products, the amorphous part is lost, and lamella crystals remain during surface erosion [10]. The same surface erosion occurs in an alkaline aqueous solution [11]. The hydrolysis reaction is unlikely to occur inside the crystal.

On the other hand, the hydrolysis of PLA in a neutral solution proceeds with a bulk erosion mechanism, in which water molecules diffuse inside the molded product. The permeability coefficient of water vapor at 25 °C decreases by half as the crystallinity of PLA increases [12]. However, with the hydrolysis of crystallized PLA in a 37 °C buffer solution, little depends on the differences in crystallinity [13]. In some cases, the PLA with greater crystallinity decomposes faster than the amorphous PLA [14], as there could be a different factor than the permeability coefficient that could be measured over a shorter time.

The development of marine-degradable materials requires accelerated testing of durability such as hydrolysis to shorten the evaluation time. Generally, the Arrhenius plot estimates the durability of polymer materials [15]. PLA crystallizes rapidly at Tg and above the temperature in hydrolysis tests [16], although crystallization progresses gradually near room temperature [13]. It is important in accelerated tests that the hydrolysis of molecular chains and changes in the crystal structure are accelerated similarly. In this study, we investigated the effect of the initial crystal structure on the hydrolysis of PLA in a high-temperature and high-humidity environment. In particular, the relationships between the internal morphology, crystallinity, and mechanical properties of molded PLA products were analyzed at different annealing temperatures and levels of humidity.

## 2. Materials and Methods

A Total Corbion (Rayong, Thailand) L175 with a melt mass flow rate (MFR) of 3 g/10 min at 190 °C was used as the PLA. The film samples were prepared using a compression-molding machine manufactured by Imoto Seisakusho (Kyoto, Japan). The PLA pellets were pressurized at 10 MPa for 2 min after heating and melting at 190 °C for 4 min under reduced pressure; then, they were cooled at 20 °C with a pressure of 10 MPa for 2 min. The film thickness was about 150 μm. To change the crystal structure, The PLA films were annealed at a predetermined temperature for 2 h in an AS ONE (Osaka, Japan) AW-250N vacuum oven. The constant temperature and humidity treatment of the sample were carried out at 70 °C and 95% relative humidity (RH) for a predetermined time using Tokyo Rika Kikai (Tokyo, Japan) KCL-2000W.

The tensile test was carried out using a Sanko (Nagoya, Japan) ISL-T300 at a span of 18 mm and a tensile speed of 1 mm/min at room temperature. The dumbbell sample was prepared by punching a JIS No. 7 dumbbell from a PLA sheet at room temperature. The crystal morphology inside the sheet was observed using an Olympus (Tokyo, Japan) BX-51P polarized optical microscope. A 20 μm-thick film was cut from the cross-section of the sheet using a sliding microtome Leica (Tokyo, Japan) RM2125. An impregnating solution was added to the preparation.

The crystal structure of the PLA was analyzed by wide-angle X-ray diffraction (WAXD) using Rigaku (Tokyo, Japan) Smart Lab. The measurement conditions were as follows: X-ray, Cu (Kα); camera length, 28.5 mm; exposure time, 10 min; and the detector was a HyPix-3000. Three samples were sandwiched between the Kapton tapes (3M Japan, Tokyo, Japan) for measurement. From the measured 2D image, the background including Kapton tape was corrected using SmartLab Studio II software (Rigaku, Tokyo, Japan). Figure 1 shows how the crystallinity was calculated using Igor Pro 6.0 software (WaveMetrics,

Portland, OR, USA). The integral value of the amorphous halo (A) in the azimuth direction was fitted by a Gaussian function to obtain an amorphous 1D profile. From the 1D profile of the semi-crystalline sample (B), the crystal planes (200) (110) and (113) (203) were fitted by a Gaussian function to obtain the area ratio of crystals to an amorphous halo.

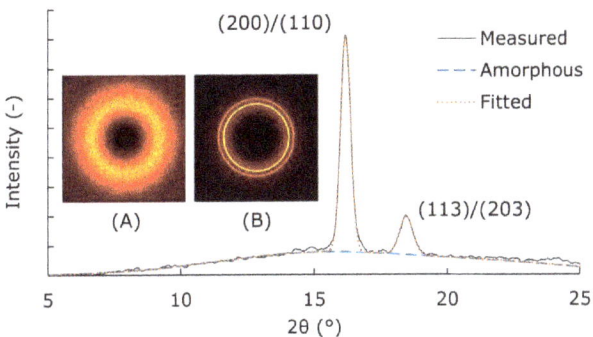

**Figure 1.** One-dimensional (1D) profile of wide-angle X-ray diffraction of the annealed PLA at 110 °C under reduced pressure. Amorphous halo (**A**); diffraction from (200), (110), (203), and (113) planes (**B**).

Thermal characterization was performed by differential scanning calorimetry (DSC) using the TA Instruments (New Castle, DE, USA) Q200. The samples were kept at room temperature for 24 h under reduced pressure before DSC measurement. The temperature was raised from 0 to 200 °C at 10 °C/min and the glass transition, cold crystallization, and melting behavior were measured. The crystallinity was calculated with the heat of fusion of the perfect crystal as 143 J/g [17]. The mobile amorphous portion was quantified from the ratio of the change in specific heat ($\Delta C_p$) at the glass transition temperature to the completely amorphous $\Delta C_p$ 0.531 J g$^{-1}$K$^{-1}$ [18]. Figure 2 shows the change in specific heat near Tg. The tangent line was extended to Tg from the specific heat of the glass state on the low-temperature side and the rubber state on the high-temperature side, and the difference was defined as $\Delta C_p$.

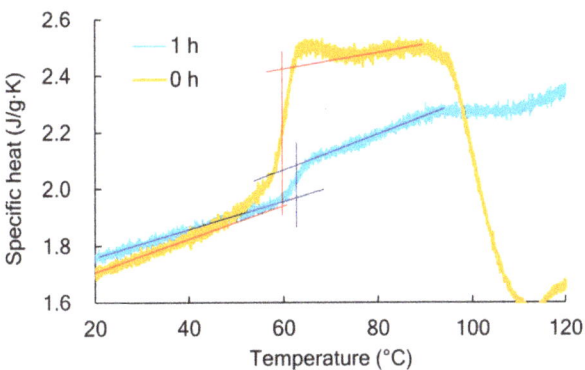

**Figure 2.** Comparison of specific heat between the amorphous PLA (0 h) and annealed PLA at 70 °C and 95% RH (1 h).

The MW of the PLA was determined by a gel permeation chromatography (GPC) using a Waters Corporation 515 HPCL pump, 2414 RI, and Agilent Technologies PLgel 5μ MID XED-C columns. The solvent was chloroform. The measurement temperature was 40 °C and the PS conversion method was used for the MW calibration. The number average molecular weight (Mn) of Total Corbion L175 was 99,600 Da, and the weight average molecular weight (Mw) was 217,000 Da. This measurement is comparable to the previously reported results [19].

## 3. Results

### 3.1. Crystal Structure of Annealed PLA

In this study, we observed changes in the crystal structure that occurred during the test at 70 °C and 95% RH using samples with varied initial crystallinities. To change the crystal structure, the compression-molded PLA film was annealed under predetermined conditions. Table 1 shows the annealing temperature, crystallinity ($\chi_c$), and tensile properties. The reference sample (STD) rapidly cooled at 20 °C was confirmed by wide-angle X-ray diffraction to be in an amorphous state with no out-of-plane orientation (Figure 1A). This STD was annealed at a predetermined temperature for 2 h under reduced pressure. Although annealing was performed at a temperature higher than the Tg of 65 °C, almost no crystallization occurred at 80 °C. The crystallinity was approximately 30% for the samples annealed from 90 to 110 °C. The crystallinity increased with the annealing, and the tensile strain at break decreased.

**Table 1.** Characteristics of PLA annealed at predetermined temperatures.

| | Annealing | $\chi_c$ * | Tensile Properties | | |
|---|---|---|---|---|---|
| | Temperature | | Modulus | Strength | Strain at Break |
| | (°C) | (%) | (MPa) | (MPa) | (%) |
| STD | – | – | 2730 | 57.9 | >50 |
| A80 | 80 | 1 | 3370 | 69.0 | >50 |
| A90 | 90 | 30 | 3270 | 71.0 | >50 |
| A100 | 100 | 32 | 3520 | 69.6 | 28 |
| A110 | 110 | 29 | 3300 | 68.0 | 5 |

* DSC method.

Annealing PLA changes not only the crystallinity but also the polymorphism. A DSC curve measured in a heating process included the information of rigid amorphous, $\alpha$-form crystals, and $\alpha'$-form crystals. In Figure 3, the STD and A80 showed a Tg around 60 °C, and the heat of cold crystallization (Tcc) around 120 °C. The peak of Tcc was apparently shifted to a lower temperature by annealing [20]. For A90, A100, and A110, a transition from $\alpha'$-form to $\alpha$-form crystal appeared at around 160 °C instead of the Tcc [21,22]. Thus, the amorphous molecular chains changed to an $\alpha'$-form crystal during annealing at less than 110 °C under reduced pressure.

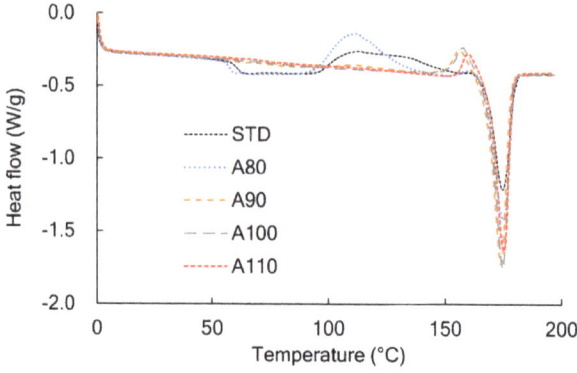

**Figure 3.** DSC curves during heating at 10 °C/min for the annealed PLA.

### 3.2. Influence of Accelerated Environment on Crystal Structure

Annealed PLA specimens were placed in a chamber at 70 °C and 95% RH for a specific length of time. Then, changes in the crystal structure were measured with POM and DSC. Figure 4 shows the POM micrographs of the cross-section of the specimens after a predetermined time had passed. Specimens at 0 h, which meant initially annealed, changed significantly depending on whether the annealing temperature was 80 °C or less or 90 °C or more. Birefringence was observed in the cross-section of the STD; the molecular chains were in-plane oriented despite the amorphous scattering by WAXD. The annealed specimens A90, A100, and A110 showed oriented crystals and spherulites. This corresponded to the crystallinity shown in Table 1.

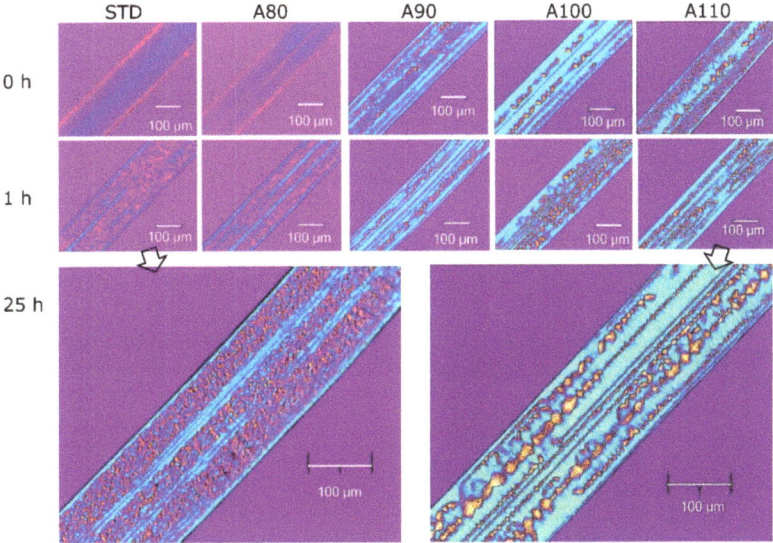

**Figure 4.** Changes in crystal morphology observed with POM during hydrolysis at 70 °C and 95% RH for PLA annealed at various temperatures.

After 1 h at 70 °C and 95% RH, the STD and A80 changed from amorphous to spherulite structures. On the other hand, the crystal morphology produced by annealing had little change at the same conditions. Even after 25 h in the chamber, there was little visual change in the crystal morphology. Thus, the initially amorphous PLA crystallized within 1 h. The crystallization rate of the hydrated PLA was higher than that in the dry state due to the increased level of molecular mobility [23]. The annealed PLA did not show the same morphology as the STD after 25 h—that is, the spherulitic morphology was formed from the amorphous PLA and not from the initially crystallized specimens. The molecular chains incorporated into the crystals indicate that they could not move freely even when hydrated.

Figure 5 shows the crystallinity measured by the DSC to quantitatively understand the change in the crystal structure. The horizontal axis represents the time for the accelerated test at 70 °C and 95% RH. The initially amorphous PLA reached a crystallinity of 30% in one hour. After 1 h or more, the crystallinity was similar regardless of the initial crystal structure. The crystal morphology shown in Figure 4 was different in the initial annealing conditions even though the crystallinity was similar. Although crystallization within 1 h is important, it was difficult to do the accelerated test in a short time due to a transient time of approximately 5 min for stabilizing the chamber. The time it takes for amorphous PLA to reach a crystallinity of 30% might be even shorter.

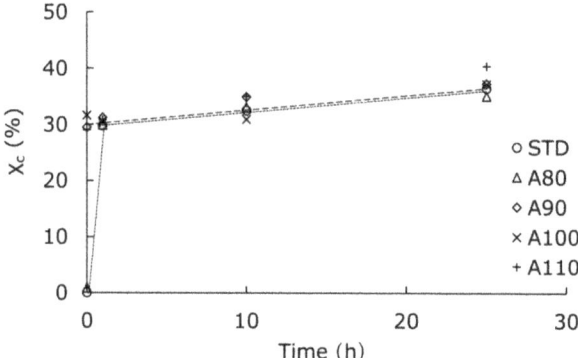

**Figure 5.** Changes in crystallinity during hydrolysis at 70 °C and 95% RH for PLA annealed at various temperatures.

### 3.3. Changes in Tensile Fracture after the Accelerated Test

Generally, crystal morphology is considered to affect the physical properties of specimens. In this study, we focused on the change in strain at breaking point in the tensile tests. Figure 6 shows the relationship between crystallinity and strain at break. Regardless of the initial crystal state, when the crystallinity reached 30% after the accelerated test, the strain at break sharply decreased. Increasing crystallinity generally reduces the strain at break. However, even with similar crystallinity, some stretched and some fractured. Thus, the critical reason for the difference in elongation was examined.

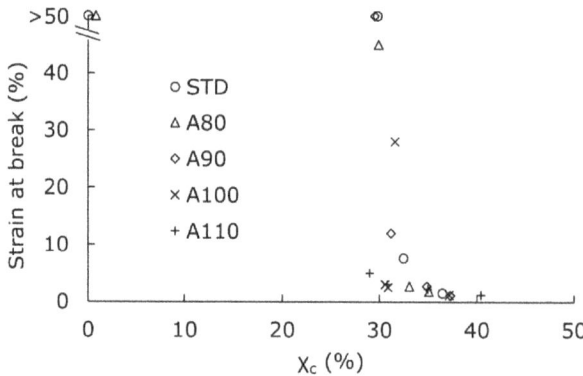

**Figure 6.** The relationship between crystallinity and strain at break for PLA annealed and hydrolyzed in predetermined conditions.

The index of higher-order structure includes not only crystallinity but also rigid amorphous. Although the crystallinity measured using DSC was the same in the specimens set for 1 h at 70 °C and 95% RH, the detailed structures were analyzed again. The WAXD analysis was added, and we attempted to separate the amorphous sections into mobile amorphous ($\chi_{maf}$) and rigid amorphous ($\chi_{raf}$) groups using the gap in specific heat change at Tg. Figure 7 shows the ratio of $\chi_c$ $\chi_{maf}$ $\chi_{raf}$ for each specimen. The comparison of the STD to the A110 suggested that the initial annealing increased $\chi_c$ and decreased $\chi_{maf}$. There was little change in $\chi_{raf}$. Thus, we focused on the amount of $\chi_{maf}$. As shown in Figure 8, strain at break increased when $\chi_{maf}$ was 17% or more. Since the tensile test was performed at a room temperature lower than Tg, the amount of $\chi_{maf}$, which indicated higher molecular mobility, affected the elongation of specimens.

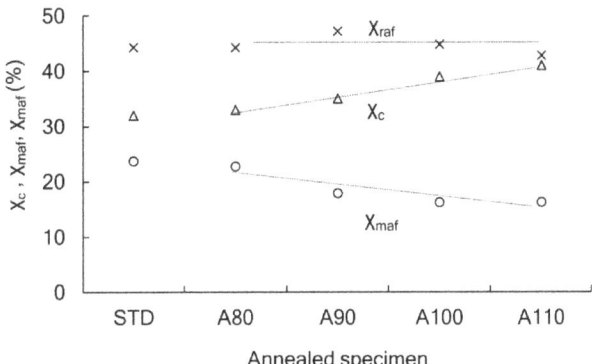

**Figure 7.** Ratio of crystal ($\chi_c$), mobile amorphous ($\chi_{maf}$), and rigid amorphous ($\chi_{raf}$) for annealed PLA after the 1 h hydrolysis at 70 °C and 95% RH.

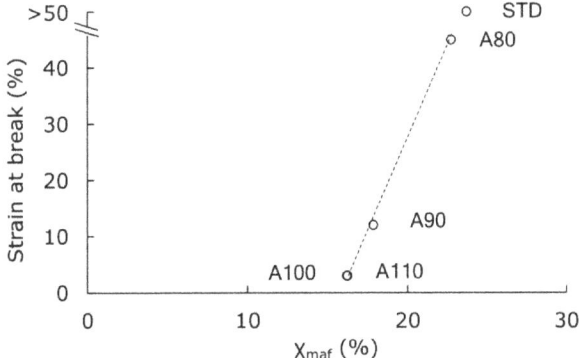

**Figure 8.** The relationship between mobile amorphous ($\chi_{maf}$) and strain at break for annealed PLA after the 1 h hydrolysis at 70 °C and 95% RH.

### 3.4. Changes in Molecular Weight Distribution after Accelerated Test

As described above, each specimen had comparable crystallinity except during the first hour. We examined whether the initial crystal structure affected hydrolysis for the remaining 24 h. Figure 9 shows the changes in the molecular weight distribution (MWD) of PLA due to hydrolysis. The MWD of PLA after the 25 h test was compared to that of raw pellets with no processing history as a reference. The Mn decreased from 217 kDa in Pellet to 98.1 kDa in STD 25 h and 122 kDa in A110 25 h. The polydispersity was 2.2, 3.1, and 2.2, respectively. The difference in the polydispersity was that the MW peak of A110 25 h shifted toward a lower MW, as shown in both curves of weight fraction w(M) and mole fraction n(M). On the other hand, in STD 25 h, the low-molecular-weight component increased remarkably.

In the decomposition of polymer chains, the asymptotic value of polydispersity is two as the secession probability of each repeating unit is equal—that is, random secession occurs [24]. Therefore, the molecular chains were randomly cleaved, although A110 was in a semi-crystalline state, where the mobility of chains was considered to be different in amorphous and crystal phases. On the other hand, the STD indicated that the low MW portion had become even lower in MWD.

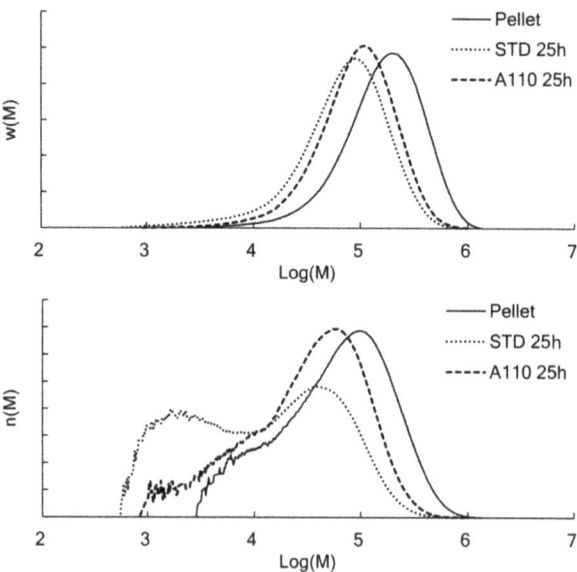

**Figure 9.** GPC curves with vertical axes of weight fraction (w(M)) and molar fraction (n(W)). Comparison of the raw pellet and the amorphous and annealed specimens hydrolyzed for 25 h.

## 4. Discussion and Conclusions

The hydrolysis of PLA often proceeds at temperatures below its Tg in the natural environment. In contrast, accelerated degradation at a high temperature and humidity is generally performed in the laboratory. In researching the relationship between the crystal structure of molded PLA products and their hydrolysis, there was concern that the difference in crystal structure between the initial specimens would be lost during the accelerated test. In this experiment as well, the amorphous PLA reached a crystallinity of 30% within 1 h at 70 °C and 95% RH—that is, the difference in crystallinity of the initially annealed specimens disappeared. However, as shown in Figure 9, the initial crystal structure greatly affected the decrease in MW.

There are two possible reasons for the difference in the MWD change by hydrolysis between STD and A110. First, the STD was hydrolyzed during the crystallizing amorphous process. Since hydrolysis is likely to proceed in the amorphous state, some molecular chains were cleaved before the crystals were sufficiently grown within 1 h at 70 °C and 95% RH, as shown in Figure 5. The low MW portion was further reduced in MW as a catalytic effect of the terminal carboxyl group of the cleaved molecular chains being concentrated in the amorphous region in the process of crystallization.

Second, the terminal groups and impurities, which promote the decomposition of polymer chains, are not as concentrated in the amorphous region in the process of A110 crystallization under reduced pressure. As shown in Figure 4, the crystal morphology of A110 is different to that of the STD crystallized in the presence of water. The STD has a fine spherulite structure due to its high molecular mobility with moisture. Thus, the diffusion of terminal groups and impurities is different in the moisture of crystallization conditions.

The hierarchy of these two reasons for accelerated hydrolysis in STD is still unknown—that is why the low MW portion is further reduced in MW. In contrast, the random decrease in MW shown by A110 indicates that it is possible to produce an industrial product that decomposes without distinguishing between crystalline and amorphous portions. Highly crystalline PLA, which has a high stiffness and thermal stability, might hydrolytically degrade without becoming microplastics made of lamellar crystals if the crystallization process is controlled. This is an industrially important finding.

**Author Contributions:** Conceptualization, H.I.; methodology, Y.K. and T.U.; formal analysis, Y.K. and T.U.; writing—original draft preparation, Y.K.; writing—review and editing, A.I. and H.I.; supervision, H.I. All authors have read and agreed to the published version of the manuscript.

**Funding:** This work is based on results obtained from a project, JPNP18016, commissioned by the New Energy and Industrial Technology Development Organization (NEDO).

**Institutional Review Board Statement:** Not applicable.

**Informed Consent Statement:** Not applicable.

**Data Availability Statement:** The data presented in this study are available on request from the corresponding author.

**Acknowledgments:** We would like to thank the work of members of our laboratory.

**Conflicts of Interest:** The authors declare no conflict of interest.

## References

1. Garlotta, D. A literature review of poly(lactic acid). *J. Polym. Environ.* **2001**, *9*, 63–84. [CrossRef]
2. Pasquini, N. *Polypropylene Handbook*, 2nd ed.; Hanser Gardner: Cincinnati, OH, USA, 2005; ISBN 9781569903858.
3. European Bioplastics BIOPLASTICS MARKET DEVELOPMENT UPDATE 2020. Available online: https://www.european-bioplastics.org/market-update-2020-bioplastics-continue-to-become-mainstream-as-the-global-bioplastics-market-is-set-to-grow-by-36-percent-over-the-next-5-years/ (accessed on 18 October 2021).
4. Galiè, F. Global Polypropylene Outlook. Available online: https://www.icis.com/explore/resources/icis-global-petrochemicals-outlook-seminar/ (accessed on 18 October 2021).
5. Matsutani, K.; Kimura, Y. Present Situation and Future Perspectives of Poly(lactic acid). In *Synthesis, Structure and Properties of Poly(lactic acid)*; Di Lorenzo, M.L., Androsch, R., Eds.; Springer International Publishing: Cham, Switzerland, 2018.
6. Choe, S.; Kim, Y.; Won, Y.; Myung, J. Bridging Three Gaps in Biodegradable Plastics: Misconceptions and Truths about Biodegradation. *Front. Chem.* **2021**, *9*, 671750. [CrossRef]
7. Lončarski, M.; Gvoić, V.; Prica, M.; Cveticanin, L.; Agbaba, J.; Tubić, A. Sorption behavior of polycyclic aromatic hydrocarbons on biodegradable polylactic acid and various nondegradable microplastics: Model fitting and mechanism analysis. *Sci. Total Environ.* **2021**, *785*, 147289. [CrossRef]
8. Manfra, L.; Marengo, V.; Libralato, G.; Costantini, M.; De Falco, F.; Cocca, M. Biodegradable polymers: A real opportunity to solve marine plastic pollution? *J. Hazard. Mater.* **2021**, *416*, 125763. [CrossRef]
9. Min, K.; Cuiffi, J.D.; Mathers, R.T. Ranking environmental degradation trends of plastic marine debris based on physical properties and molecular structure. *Nat. Commun.* **2020**, *11*, 727. [CrossRef]
10. MacDonald, R.T.; McCarthy, S.P.; Gross, R.A. Enzymatic degradability of poly(lactide): Effects of chain stereochemistry and material crystallinity. *Macromolecules* **1996**, *29*, 7356–7361. [CrossRef]
11. Tsuji, H.; Ikada, Y. Properties and morphology of poly(L-lactide). II. Hydrolysis in alkaline solution. *J. Polym. Sci. Part A Polym. Chem.* **1998**, *36*, 59–66. [CrossRef]
12. Tsuji, H.; Tsuruno, T. Water vapor permeability of poly(L-lactide)/poly(D-lactide) stereocomplexes. *Macromol. Mater. Eng.* **2010**, *295*, 709–715. [CrossRef]
13. Tsuji, H.; Ikada, Y. Properties and morphology of poly(L-lactide) 4. Effects of structural parameters on long-term hydrolysis of poly(L-lactide) in phosphate-buffered solution. *Polym. Degrad. Stab.* **2000**, *67*, 179–189. [CrossRef]
14. Tsuji, H.; Mizuno, A.; Ikada, Y. Properties and morphology of poly(L-lactide). III. Effects of initial crystallinity on long-term in vitro hydrolysis of high molecular weight poly(L-lactide) film in phosphate-buffered solution. *J. Appl. Polym. Sci.* **2000**, *77*, 1452–1464. [CrossRef]
15. Brown, R.P. Predictive techniques and models for durability tests. *Polym. Test.* **1995**, *14*, 403–414. [CrossRef]
16. Zhang, X.; Espiritu, M.; Bilyk, A.; Kurniawan, L. Morphological behaviour of poly(lactic acid) during hydrolytic degradation. *Polym. Degrad. Stab.* **2008**, *93*, 1964–1970. [CrossRef]
17. Righetti, M.C.; Gazzano, M.; Di Lorenzo, M.L.; Androsch, R. Enthalpy of melting of $\alpha'$- and $\alpha$-crystals of poly(L-lactic acid). *Eur. Polym. J.* **2015**, *70*, 215–220. [CrossRef]
18. Henricks, J.; Boyum, M.; Zheng, W. Crystallization kinetics and structure evolution of a polylactic acid during melt and cold crystallization. *J. Therm. Anal. Calorim.* **2015**, *120*, 1765–1774. [CrossRef]
19. Srithep, Y.; Pholharn, D.; Akkaprasa, T. Effect of molecular weight of poly(L-lactic acid) on the stereocomplex formation between enantiomeric poly(lactic acid)s blendings. In Proceedings of the IOP Conference Series: Materials Science and Engineering, Beijing, China, 19–22 August 2019; IOP Publishing: Bristol, UK, 2019; Volume 526, p. 012024.
20. Srithep, Y.; Nealey, P.; Turng, L.S. Effects of annealing time and temperature on the crystallinity and heat resistance behavior of injection-molded poly(lactic acid). *Polym. Eng. Sci.* **2013**, *53*, 580–588. [CrossRef]
21. Androsch, R.; Schick, C.; Di Lorenzo, M.L. Melting of conformationally disordered crystals ($\alpha'$ phase) of poly(l-lactic acid). *Macromol. Chem. Phys.* **2014**, *215*, 1134–1139. [CrossRef]

22. Di Lorenzo, M.L.; Androsch, R. Influence of $\alpha'$-/$\alpha$-crystal polymorphism on properties of poly(l-lactic acid). *Polym. Int.* **2019**, *68*, 320–334. [CrossRef]

23. Vyavahare, O.; Ng, D.; Hsu, S.L. Analysis of structural rearrangements of poly(lactic acid) in the presence of water. *J. Phys. Chem. B* **2014**, *118*, 4185–4193. [CrossRef] [PubMed]

24. Schnabel, W. *Polymer Degradation: Principles and Practical applications*; Carl Hanser Verlag: Munich, Germany, 1981; pp. 15–19. ISBN 3446132643.

*Article*

# Fully Recyclable Bio-Based Epoxy Formulations Using Epoxidized Precursors from Waste Flour: Thermal and Mechanical Characterization

Francesca Ferrari [1], Carola Esposito Corcione [1,*], Raffaella Striani [1], Lorena Saitta [2], Gianluca Cicala [2] and Antonio Greco [1]

[1] Department of Engineering for Innovation, University of Salento, Via Arnesano, 73100 Lecce, Italy; francesca.ferrari@unisalento.it (F.F.); raffaella.striani@unisalento.it (R.S.); antonio.greco@unisalento.it (A.G.)
[2] Department of Civil Engineering and Architecture (DICAR), University of Catania, Viale Andrea Doria 6, 95125 Catania, Italy; lorena.saitta@phd.unict.it (L.S.); gianluca.cicala@unict.it (G.C.)
* Correspondence: carola.corcione@unisalento.it

**Abstract:** Organic wastes represent an increasing pollution problem due to the exponential growth of their presence in the waste stream. Among these, waste flour cannot be easily reused by transforming it into high-value-added products. Another major problem is represented by epoxy-based thermosets, which have wide use but also poor recyclability. The object of the present paper is, therefore, to analyze both of these problems and come up with innovative solutions. Indeed, we propose a completely new approach, aimed at reusing the organic waste flour, by converting it into high-value epoxy-based thermosets that could be fully recycled into a reusable plastic matrix when added to the waste epoxy-based thermosets. Throughout the research activity, the organic waste was transformed into an epoxidized prepolymer, which was then mixed with a bio-based monomer cured with a cleavable ammine. The latter reactant was based on Recyclamine™ by Connora Technologies, and in this paper, we demonstrate that this original approach could work with the synthetized epoxy prepolymers derived from the waste flour. The cured epoxies were fully characterized in terms of their thermal, rheological, and flexural properties. The results obtained showed optimal recyclability of the new resin developed.

**Keywords:** epoxidation; thermoset recycling; organic waste

**Citation:** Ferrari, F.; Esposito Corcione, C.; Striani, R.; Saitta, L.; Cicala, G.; Greco, A. Fully Recyclable Bio-Based Epoxy Formulations Using Epoxidized Precursors from Waste Flour: Thermal and Mechanical Characterization. *Polymers* **2021**, *13*, 2768. https://doi.org/10.3390/polym13162768

Academic Editor: Piotr Bulak

Received: 26 July 2021
Accepted: 12 August 2021
Published: 18 August 2021

**Publisher's Note:** MDPI stays neutral with regard to jurisdictional claims in published maps and institutional affiliations.

## 1. Introduction

The use of epoxy-based composites is widely accepted in different fields. In the aerospace sector, epoxy resins are used because of its low cost and suitability for producing large structures. Recent studies reported novel technologies for producing enhanced composites for the aeronautical field. Zotti et al. [1] developed PDA-coated silica nanoparticles as filler for a common aeronautical epoxy resin, improving the mechanical properties, the damage resistance, and the thermal stability with respect to the neat matrix. Liquid resin infusion of epoxy resins is well established in the transportation and naval sectors. In the civil sector, the use of epoxy composites is widely accepted for semistructural and structural applications. An increasing interest was recently devoted to the epoxy–timber composites in the construction field. Awad et al. [2] studied the effect of calcium sulfate as a UV absorber able to improve the aging of two cured epoxies. However, increased awareness of the environmental impact of thermosets has raised concerns regarding their use and has pressed the industry and academia to develop tailored recycling strategies for epoxy-based composites. An additional environmental limitation of currently used epoxy systems is the use of petroleum-based raw materials for their synthesis. Life cycle analysis (LCA) must be considered to develop resins and composites complying with the cradle-to-cradle strategy [3].

Rybicka et al. [4] described the technology readiness level (TRL) of several recycling technologies: Incineration and landfilling were classified as TRL 9; pyrolysis for carbon and glass fiber composites resulted in a TRL 8. The fluidized bed pyrolysis and solvolysis process achieved a median TRL of 4. In some recent reviews, the annual capacity of several technologies for recycling carbon-fiber-reinforced composites was discussed [5,6]. Pyrolysis was confirmed as the approach reaching capacities in the range of 1000–2000 tons. The main limitations of thermal and mechanical recycling processes are fibers' property degradation and that matrices are fully depolymerized with partial recovery in useful forms [7].

Chemical recycling is emerging as a viable approach to recover clean and undamaged reinforcing fibers while allowing for the recovery of monomers or oligomers that can be reused. Xu et al. [8] presented an approach based on the decomposition of epoxy-based composites in a $H_2O_2$/acetone mixed solution heated between 60 and 150 °C. The degraded products were analyzed, showing mainly bisphenol A and its derivatives, such as phenol derivatives, which are generated during the decomposition of the epoxy network. Wang et al. [9] developed a recycling approach based on the use of acetic acid to swell the composites while the weakly coordinating aluminum ions in $CH_3COOH$ solution selectively cleaved the C−N bond, allowing for obtaining oligomers from the epoxy resins. These papers demonstrated the possibility to cleave epoxy networks, but the reuse strategies for the recovered oligomers were not assessed.

Back in 2012, the company Connora Technologies presented a novel class of ammine reagents named Recyclamine™, designed to be selectively cleaved in an aqueous solution with acetic acid, using mild conditions (i.e., 80 °C), thus yielding clean reinforcing fibers and a reusable thermoplastic matrix from the epoxy network. The Recyclamine™ reactants were characterized in terms of their aging resistance [10], and mixing them with bio-based epoxy monomers, their properties using high-pressure resin transfer molding [11] and resin infusion [12] were measured. The recycling process of bio-based resin cured by Recyclamine™ was investigated by life cycle analysis (LCA), confirming its potential to offer a disruptive solution to the end-of-life problem of epoxy-based composites [10,13,14]. In a recent study, the benefits of Recyclamine™ in terms of life cycle costing (LCC) were assessed [13].

The use of petroleum-based raw materials for the synthesis of epoxy monomers is another limit of the epoxy-based composites used nowadays. To overcome this limit, several researchers developed bio-based epoxy monomers synthetized from vegetable oils [14], natural acids [15,16], lignin [17], and so forth. The procedures for the synthesis of epoxy precursors from natural renewable resources require, in most of the cases, the use of organic solvents limiting the development of a truly green approach. In a recent paper, Esposito Corcione et al. [18] presented an innovative approach to obtain epoxidized monomers starting from waste flours recovered from the processing waste of pasta factories or from the organic fraction of municipal solid waste. This approach simply relies on waste's treatment with UV/ozone radiations without the use of any solvents. This treatment is fast, cheap, reliable, and with no toxic emissions. The amount of municipal solid waste globally collected per year is approximately $1.3 \times 10^{12}$ t, and it is expected to rise up to $2.2 \times 10^{12}$ t per year by 2025 (EPA (United States Environmental Protection Agency), 2017).

The huge amount of organic waste is becoming an increasing issue for the waste management of modern cities, while the technology developed by Esposito Corcione et al. [18,19] can turn waste into a high-value-added product. However, avoiding global negative impacts when using epoxy monomers is important to develop suitable recycling approaches to reuse epoxy resins at their end of life. This approach respects the cradle-to-cradle strategy.

In the present paper, epoxidized monomers synthetized from waste flour were mixed with bio-based epoxy precursors; then the epoxy blends obtained were cured using a cleavable ammine to develop a fully recyclable bio-based epoxy thermoset. The resins were fully characterized in terms of thermal and mechanical properties to optimize their final performances. The optimized formulation was recycled using only an acidic aqueous solution under mild conditions to demonstrate the possibility to recover a reusable plastic from the cured epoxy resin.

## 2. Materials and Methods

Polar Bear (R-Concept, Barcelona, Spain), a bio-based epoxy system designed specifically for the composite processing.

Recyclamine R-101 (R-Concept, Barcelona, Spain), is a recyclable epoxy cure agent for composite manufacturing. Polar Bear and Recyclamine R-101 are both liquid at room temperature.

Waste flour (WF) was obtained from the processing waste of the pasta factories. Epoxidized waste flour (EWF) was obtained by contemporary exposure to UV radiations and ozone for 5 h following the method reported in a previous work [18] and in a patent application [20]. A medium-pressure Hg UV lamp (UV HG 200 ULTRA, Jelosil Srl, Vimodrone, Italy), with a radiation intensity on the surface of the samples of 9.60 $W/mm^2$, was used for waste flour treatment.

FTIR analysis, performed with an FTIR (6300 Spectrometer, Jasco, Cremella, Italy), was used to assess the presence of epoxy groups after UV/ozone exposure. Infrared spectra were recorded in the wavelength range between 400 and 4000 $cm^{-1}$, 128 scans, and 4 $cm^{-1}$ of resolution by using a germanium round crystal window. The spectra acquisition was carried out before and after the UV/ozone and after curing.

As reported in our previous article [20], the epoxy content of the waste flour was checked by using titration, carried out according to Method A of ASTM D 1652-97 (ASTM D (1652)-97, 1997). The amount of the consumed acid during titration, which is an index of the epoxy content of the sample, was used to calculate the epoxy content (E) and the equivalent epoxy weight (WPE) of the waste flour.

Different samples were produced by mixing the Polar Bear resin with specific amounts of epoxidized waste flour and Recyclamine R-101. First, different blends were produced by varying the ratio between Polar Bear and waste flour, keeping a constant amine content (Table 1). The initial amount of amine was chosen by considering the value suggested from R-Concept for the blend of Polar Bear and Recyclamine R-101, corresponding to 22 phr.

**Table 1.** Samples at different commercial/epoxidized waste flour ratios and constant amine content.

| Sample Name | Polar Bear (wt%) | Epoxidized Waste Flour (wt%) | Recyclamine R-101 (phr) |
|---|---|---|---|
| P_A22 | 100 | - | 22 |
| EWF35_A22 | 65 | 35 | 22 |
| EWF40_A22 | 60 | 40 | 22 |
| EWF50_A22 | 50 | 50 | 22 |

After choosing the optimal ratio between Polar Bear and waste flour, the amine content was varied by adding different phr's of Recyclamine R-101 (Table 2) in order to optimize the curing kinetic of the system.

**Table 2.** Samples at constant commercial/epoxidized waste flour ratio and varying amine contents.

| Sample Name | Polar Bear (%) | Epoxidized Waste Flourn (%) | Recyclamine R-101 (phr) |
|---|---|---|---|
| EWF50_A15 | 50 | 50 | 15 |
| EWF50_A20 | 50 | 50 | 20 |
| EWF50_A22 | 50 | 50 | 22 |
| EWF50_A24 | 50 | 50 | 24 |
| EWF50_A26 | 50 | 50 | 26 |
| EWF50_A28 | 50 | 50 | 28 |
| EWF50_A30 | 50 | 50 | 30 |

All the mixtures were degassed by applying vacuum at room temperature, then pouring in silicon molds and curing for 24 h at room temperature. The cure was followed by a postcure process for 2 h in a static oven. The postcure temperature was varied between 120, 150, and 160 °C, as shown in Figure 1.

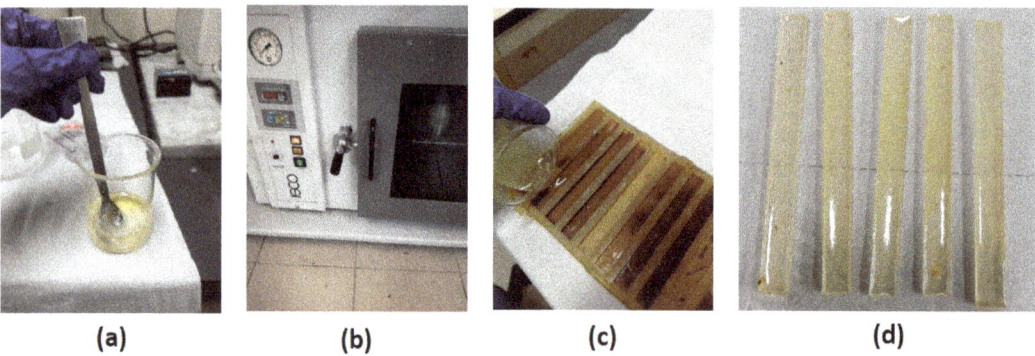

(a)          (b)          (c)          (d)

**Figure 1.** Mixing (**a**), vacuum (**b**), and pouring in silicon molds (**c**) cured thermoset samples (**d**).

### 2.1. Recycling Procedure

Recyclamine™ is an epoxy hardener developed by Connora Technologies that allows for obtaining a recyclable thermoset that can be converted into a meltable thermoplastic. Its recyclability key factor is based on the presence of amino-acid-cleavable groups that allow the cleavage of the crosslink points of the epoxy-cured network [21].

The resin system selected for the recycling trials was EWF50_A15, which showed the best properties among all the resin systems tested, as it will be shown in the paper. The chemical recycling procedure is schematically drawn in Figure 2.

**Figure 2.** Chemical recycling process's main steps.

A sample of 5 g of the epoxy system EWF50_A15 was solubilized in 300 mL of 25 %vol acid acetic solution (CH$_3$COOH) at 80 °C for 1 h. The obtained mixture was rotoevaporated at 60 °C at a pressure ranging between 110 and 60 mbar and at a rotation speed of 3500 rpm. The distilled acetic acid was stocked, being reusable for a new chemical recycling treatment, while the concentrated solution obtained (about 75 mL) was neutralized in 300 mL of 50 %vol ammonium hydroxide solution. During this phase, a whitish compound started to precipitate, which was the recycled thermoplastic of interest. Then, the solution containing the precipitate was centrifuged for 5 min at 3000 rpm and, at the end, the supernatant removed. The solid phase at the bottom of the test tube was recovered and washed in ionized water to remove any residual traces of acetic acid and ammonium hydroxide solutions. Eventually, it was dried in a vacuum stove for 24 h at 50 °C. The thermoplastic obtained is a brown compact solid shown in the panel of Figure 2. The recycling process applied previously on bio-based epoxy derived from pine oil and paper byproducts [22] resulted in a white solid. However, in this paper we modified the recycling process compared with the one used previously [21] in the following steps: the use of a Rotavapor to concentrate the solution, the replacement of sodium hydroxide with ammonium hydroxide in the neutralization phase, and the use of centrifugation in place of filtration.

The recycling process yield was equal to 85%, in the same range of the yields obtained previously [22]. However, the new process was faster and allowed for reusing the ammonium hydroxide solution, thus leading to a greener process, which is under evaluation using LCA to quantify the environmental benefits.

### 2.2. Methods

Rheological analyses were carried out with a Rheometrics Ares rheometer. A double plate geometry was used, setting a gap of 0.3 mm, constant oscillatory amplitude (1%), and frequency (1 Hz). The tests consisted of a temperature ramp from room temperature to 130 °C.

DSC analysis was performed on a Mettler Toledo 622 differential scanning calorimeter (DSC). Samples were heated from 25 to 250 °C at 20 °C/min in air.

The thermal stability of the films was assessed by TGA, with a TA Instruments SDT Q600 (TA Instruments, New Castle, DE, USA). The samples were heated in an alumina holder from 20 to 600 °C at a heating rate of 10 °C/min under air atmosphere; three measurements were performed on each sample.

The flexural properties of each cured sample were measured using a dynamometer, Lloyd LR5K, according to ASTM D790 (ASTM D790-17, 2017) (three points bending with the specimen dimension: 80 mm × 10 mm × 4 mm). Five replicates were performed on each sample.

Dynamic mechanical analysis was carried out on a dynamic mechanical thermal analyzer (TRITEC2000 by Triton Technology, Leicestershire, UK) by single cantilever geometry. The recycled polymers, after 1 day drying at 40 °C, were tested in their powder form using the pocket DMA approach, a technique used for testing powders in the pharmaceutical field [23] and for polymer blends [24] The polymers obtained from recycling were finely micronized in powder with an average dimension of 30 μm. Then 0.35 g of polymer powder was weighted in a standard stainless steel pocket purchased from Triton and pressed to obtain a uniform thickness. The test was carried out according to the following protocol: the sample was stabilized at 25 °C and then heated up to 180 °C at 5 °C min$^{-1}$; the samples were cooled down naturally and reheated up to 180 °C at 5 °C min$^{-1}$. Similar techniques were also reported by Carlier et al. [25] for organic polymers under the name supported DMA. This kind of technique allows for direct evaluation of thermal transitions from E′ and tan d traces. However, the absolute values of E′ and tan δ for the polymer are influenced by the presence of the metal pocket, and thus, the real values should be analyzed considering the assembly as a sandwich material. The tan δ versus temperature was plotted.

*2.3. Statistical Analysis*

Analysis of variance (ANOVA) was used to highlight the statistical significance of different parameters, as the different amounts of EWF and Polar resin, on the mechanic properties. For this purpose, the F value, which is defined as the ratio of the variation between sample means to the variation within the samples, was calculated from the measured data. Then, being "a" the number of levels of the variance factor and "n" the number of tests for each level, the critical F value, FCV(a-1, a(n-1), α), can be estimated. FCV represents the value of F distribution with degrees of freedom (a-1) and a(n-1), which, at a confidence level, α, corresponds to the null hypothesis (equivalence of the means). Therefore, F < FCV indicates that the population means are equivalent, whereas F > FCV indicates that the population means are significantly different. Another quantitative measure for reporting the result of a test of hypothesis is the *p*-value. The *p*-value is the probability of the test statistic to be at least as extreme as the one observed, given that the null hypothesis is true. A small *p*-value is an indication that the null hypothesis is false. It is good practice to decide in advance of the test how small a *p*-value is required to reject the test, that is, to choose a significance level, α, for the test. For example, it can be decided to reject the null hypothesis if the test statistic exceeds the critical value (for α = 0.05) or, analogously, to reject the null hypothesis if the *p*-value is smaller than 0.05.

## 3. Results and Discussion

The FTIR spectra on waste flour are reported in Figure 3; in particular, the FTIR curve of waste flour (WF) shows the typical peaks of starch: 1412 cm$^{-1}$ assigned to –CH$_2$ bending and –COO stretch, 1075, 1048 cm$^{-1}$ and 1022 cm$^{-1}$ assigned to the crystalline and amorphous regions of starch, respectively, and 1164 cm$^{-1}$ assigned to vibrations of the glucosidic C–O–C bond and the whole glucose ring that can present different modes of vibrations and bending conformations.

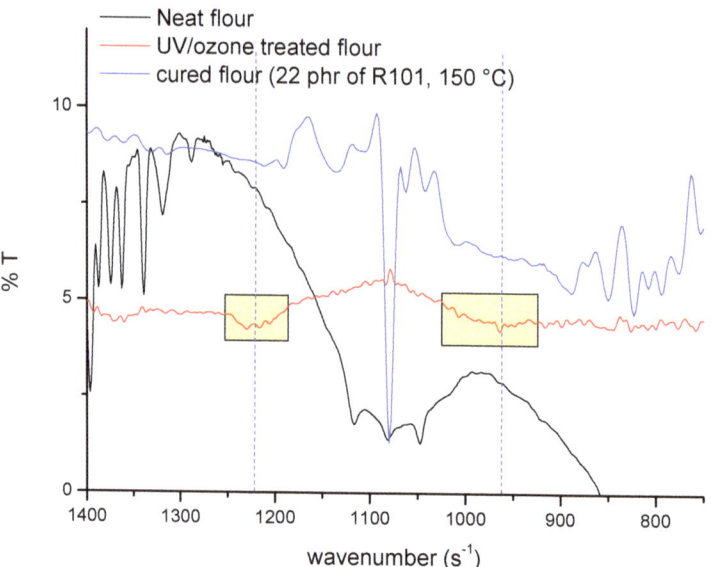

**Figure 3.** FTIR spectra of waste flour.

A contemporary exposure of waste flour to UV/ozone radiation involves the appearance of the typical signals of the epoxy ring at 1260, 890, and 827 cm$^{-1}$. This indicates that the treatment allows for obtaining epoxidized waste flour (EWF).

*Polymers* **2021**, *13*, 2768

After curing of the waste flour in the presence of the amine, the peaks due to the epoxy rings disappear, confirming epoxy curing reaction. The strong peak at 1075 cm$^{-1}$ can be again attributed to the bending vibration of residual glucose.

The viscosity curves of the commercial system and its blend with 50% of EWF are reported in Figure 4. The commercial system is characterized by a step increase in viscosity of around 95 °C, which is indicative of the reaction with the amine. The addition of waste flour involves a decrease in the onset temperature of reaction, as clearly observed in Figure 4a. This indicates that in the presence of EWF, the crosslinking reaction of the system is accelerated.

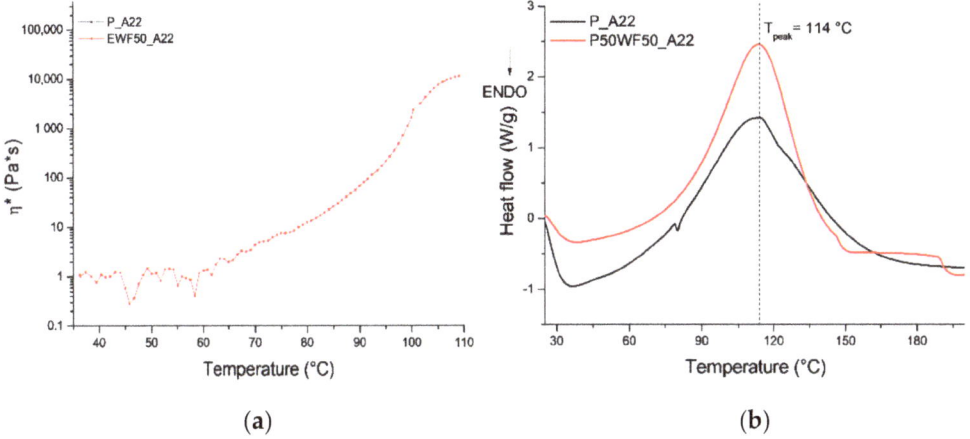

**Figure 4.** Rheological (**a**) and DSC (**b**) analysis of the curing reaction.

However, as shown in Figure 4b, the exothermal peak temperature obtained by DSC analysis is the same for both systems. A comparison between the rheological and DSC curve of the commercial system shows that viscosity increase occurs around 100 °C, where, however, the extent of reaction, as measured by DSC, is still quite low. This indicates that the rheological analysis can only provide information at a relatively low degree of conversion. The lower onset temperature of viscosity increase observed in Figure 4a for the system with EWF is therefore relative to very low conversions, where probably DSC analysis is not able to detect the very slow heat release of the reaction.

In order to choose the optimal postcure temperature, DSC scans were carried out on systems postcured at three different temperatures (120, 150, and 160 °C) after the cure at room temperature for 24 h.

The DSC curves in the temperature range between 30 and 170 °C, reported in Figure 5a–c for the cured systems, show that the postcure temperature has no significant effect on the glass transition of the commercial system, which is, in any case, around 96 °C. On the other hand, two different glass transition signals were detected in the blend with EWF, which indicates a partial miscibility of the system. Both Tg values were significantly affected by the postcure temperature. The higher values were found after postcure at 150 °C. The decrease of the Tg values after further increasing the postcure at 160 °C was due to the poor thermal stability of the EWF, which, from TGA analysis, was found to have an onset temperature of degradation of around 170 °C.

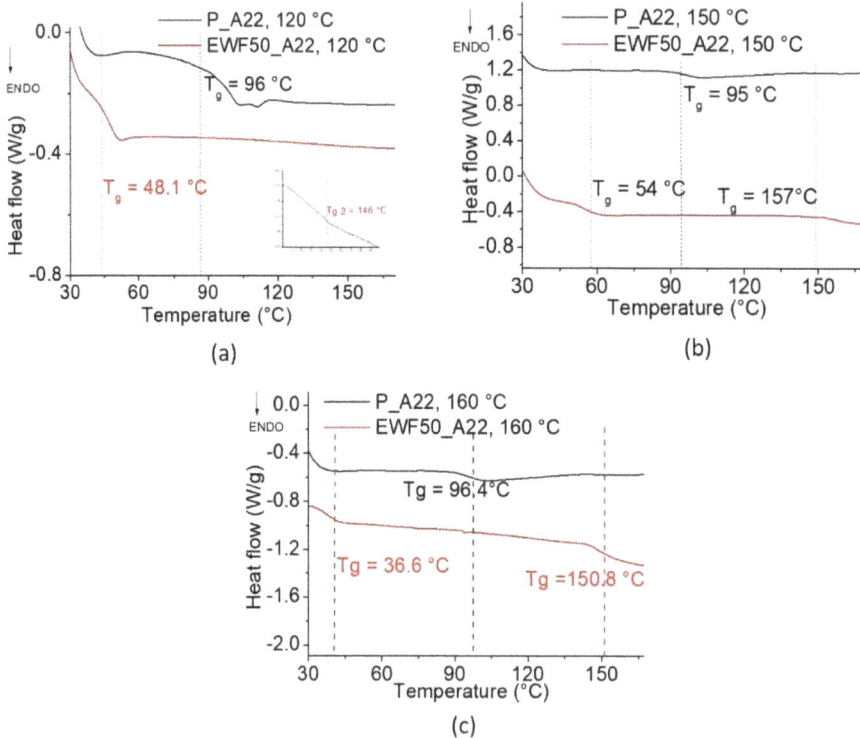

**Figure 5.** DSC analysis of blends postcured at 120 °C (**a**), 150 °C (**b**), and 160 °C (**c**).

With the aim of developing an epoxy system that could find use in different industrial applications, a glass transition at around 140–150 °C is high enough to guarantee good performances and stability of properties. On the other hand, the lower Tg represents a limit of the developed system.

Therefore, our further efforts were aimed at increasing the lower Tg signal. In the following analysis, we will only focus on the lower Tg, neglecting potential changes of the higher Tg.

Figure 6 shows DSC heating scans on samples postcured at 150 °C with different amounts of waste flour. No difference in the lower glass transition temperature was detected with increasing EWF content. Therefore, in order to increase the amount of recycled material, we focused our further analysis on the blend at 50% of EWF. Unfortunately, it was impossible to further increase the amount of EWF since this resulted in a significant increase in the liquid blend viscosity. However, the Tg value of the blends was much lower than that of the neat commercial systems, which required further optimization of the amine content in order to increase the glass transition of the system.

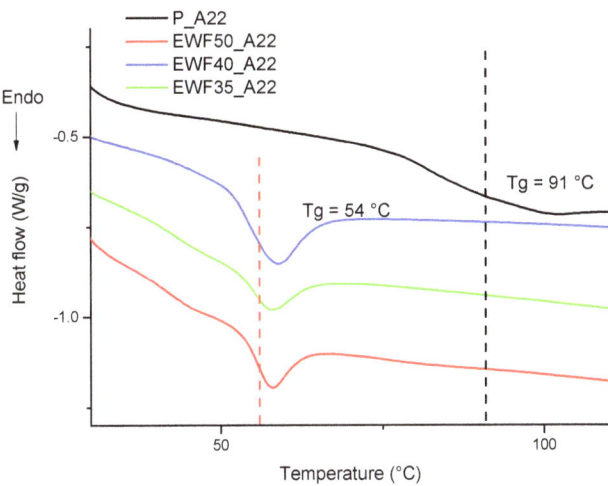

**Figure 6.** DSC analysis of postcured blends at different EWF contents.

In Figure 7, the DSC curves of samples at a constant EWF content and varying amine amounts are reported. A lower amine amount allowed for increasing the lower glass transition of the system. This indicates that when the amount of amine was too high, an excess of uncured amine remained in the sample after curing. This amine can effectively act as a plasticizer for the epoxy, significantly reducing the glass transition signal. A similar behavior was in fact observed for the commercial system at higher amine contents, as reported in Figure 8. Additionally, in this case, the excess of unreacted amine caused a plasticization effect, which reduced the Tg of the system. According to our analysis, it would be possible to further reduce the amount of amine. However, for the same reason previously discussed, the amount of amine was not further reduced below 15 phr because it resulted in very high viscosities.

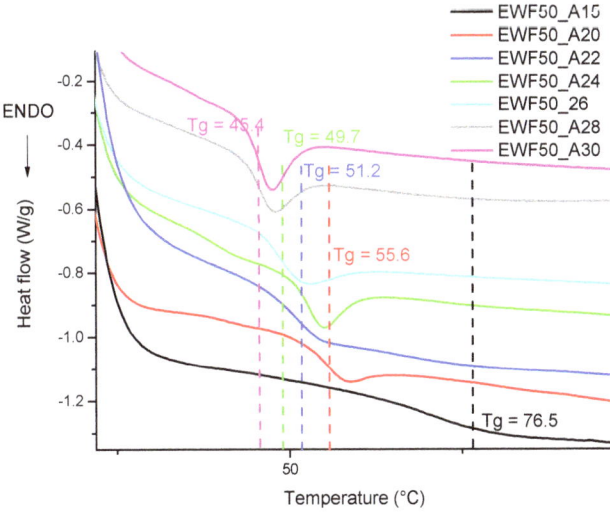

**Figure 7.** DSC analysis of postcured blends at different amine contents.

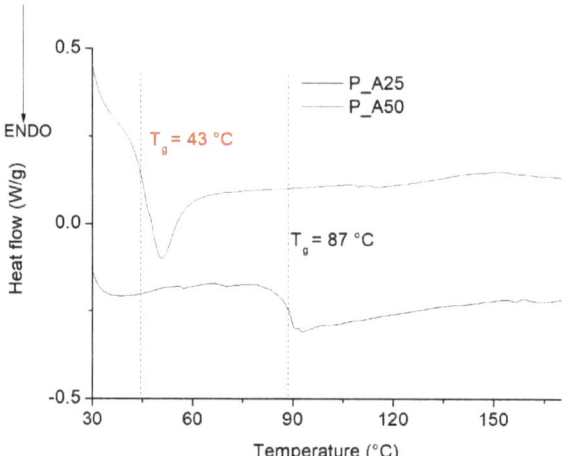

**Figure 8.** DSC analysis of postcured commercial epoxy at different amine contents.

Thermogravimetric analyses, shown in Figure 9, were performed on EWF without and with the addition of different amounts of Polar resin. The EWF sample showed a first weight loss below 150 °C, mainly attributed to the water evaporation, and a second loss between 150 and 350 °C, due to the degradation of the starch. The addition of Polar resin involved, in any case, a strong decrease in the absorption of water. Additionally, an increase in thermal stability was detected in the second stage, between 250 and 350 °C, where the production of a carbonaceous residue at lower temperatures occurred together with the degradation of the flour. In this step, a decrease in weight loss was detected by increasing the Polar resin content. The second stage was followed by a third degradation step, characterized by the oxidation of the remaining carbonaceous char; also in this step, the production of a higher final solid residue was detected with higher Polar content.

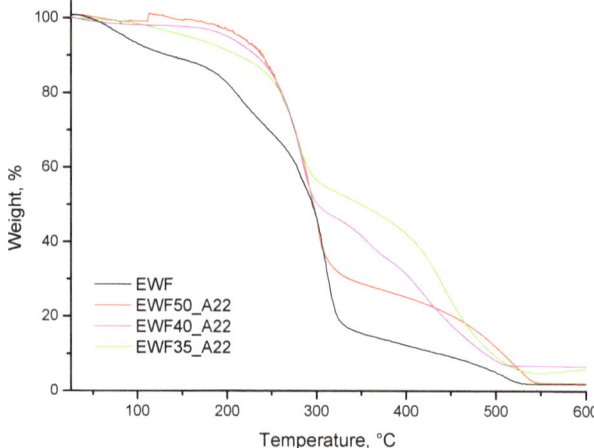

**Figure 9.** TGA on EWF with different Polar resin contents.

In Figure 10, the stress–strain curves obtained from flexural tests on samples with different EWF contents are reported. Results from flexural tests, and the mechanical properties reported in Table 3, confirmed the results from DSC analysis. The lower Tg found in Figure 6 for the blends, compared with the neat commercial system, resulted in lower mechanical properties too. The flexural modulus and strength of the blends were significantly lower compared with the commercial system. This was first due to the structure of the samples. In particular, the structure of the manufactured system was influenced by the presence of a crystalline zone of the waste flour typical of starch, which remained unchanged even after the progressive reprocessing cycles. As reported in our previous work [20], waste flour showed a semicrystalline nature, without any significant change in the crystalline fraction and the crystal planes, compared with native starch. The presence of this crystalline fraction involved an increase in brittleness of the sample, compared with the completely amorphous Polar Bear–Recyclamine system. On the other hand, the addition of Polar Bear resin involved a strong increase in the mechanical response, compared with the results found in our previous work [20] for the sample made up of only waste flour, characterized by very low flexural strength (7.32 ± 0.65 MPa). The addition of Polar resin, in fact, allowed a reduction of the high number of voids and defects occurring during both the water evaporation and the curing process of neat waste flour samples.

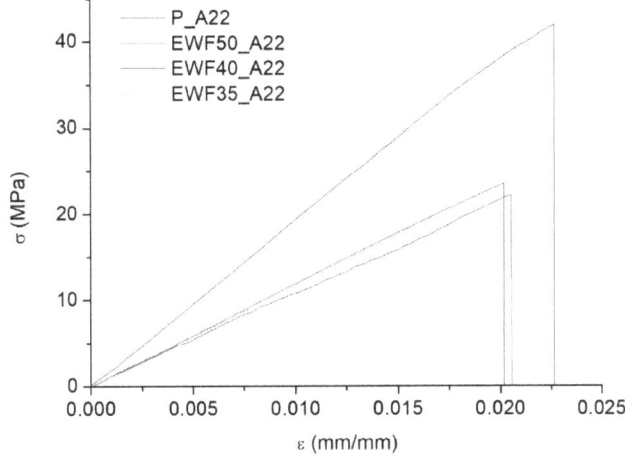

**Figure 10.** Stress–strain curves from flexural tests of cured blends at different EWF contents.

**Table 3.** Flexural properties of cured blends at different EWF contents.

| Sample | $\sigma_R$ (MPa) | $\varepsilon_R$ (mm/mm) | E (MPa) |
|---|---|---|---|
| P_A22 | 37.4 ± 11.4 | 0.021 ± 0.008 | 1893 ± 340 |
| EWF50_A22 | 20.1 ± 5.9 | 0.020 ± 0.004 | 1313 ± 430 |
| EWF40_A22 | 18.5 ± 3.5 | 0.021 ± 0.007 | 1146 ± 245 |
| EWF35_A22 | 19.1 ± 4.3 | 0.014 ± 0.009 | 1371 ± 188 |

This was confirmed by one-way analysis of variance (ANOVA). Considering the amount of EWF as the source of variation, with four levels, and three degrees of freedom, its significance on flexural modulus, strength, and strain at break was tested by calculating the F value as the ratio of the variance between the means to the variance of the experimental error. The F value was then used in order to calculate the corresponding *p*-value, which was then compared with the confidence level, $\alpha = 0.05$. According to ANOVA, $p > \alpha$ corresponds to the null hypothesis (equivalence of the means), whereas $p < \alpha$ indicates that the population means were significantly different. For flexural strength and modulus, $p = 0.0013$ and $p = 0.01$, respectively, indicate the statistically relevant effect of the addition of EWF on the corresponding property of the blend. In contrast, for strain at break, $p = 0.39$ indicates that the effect of EWF was not statistically significant.

On the other hand, limiting the ANOVA to the blends, and therefore neglecting the sample of commercial epoxy, with three levels, and two degrees of freedom, for flexural strength and modulus, $p = 0.85$ and $p = 0.50$, respectively, indicates that the corresponding mechanical properties were not influenced by the amount of EWF in the tested range of compositions. This is consistent with the fact that, in Figure 6, the lower Tg signal was independent on the amount of EWF.

In Figure 11, samples with different amine contents are compared. The mechanical properties calculated from the stress–strain curves are reported in Table 4. ANOVA was again used to establish the effect of the amount of amine on the mechanical properties of the sample. Considering the data in Table 4, neglecting the commercial sample, and therefore considering seven levels, and six degrees of freedom, for flexural strength, modulus, and strain at break, $p = 2.1 \times 10^{-11}$, $p = 8.3 \times 10^{-6}$, and $p = 8.2 \times 10^{-4}$, respectively, indicate that the amount of amine had a statistically significant effect on each of the mechanical properties. Referring to the data in Table 4, each of the mechanical properties increased with a decreasing amount of amine. The lower modulus and strength found for the higher amine content confirmed that excess amine acts as a plasticizer for the cured epoxy, which, as discussed for Figure 7, involved the Tg reduction.

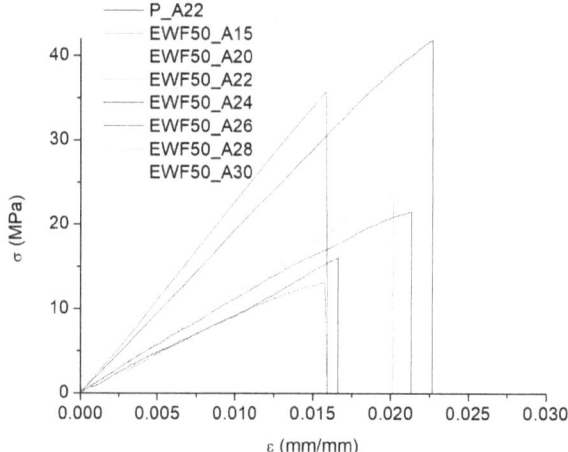

**Figure 11.** Stress–strain curves from flexural tests of cured blends at different amine contents.

**Table 4.** Flexural properties of cured blends at different amine contents.

| Sample | $\sigma_R$ (MPa) | $\varepsilon_R$ (mm/mm) | E (MPa) |
|---|---|---|---|
| P_A22 | 37.4 ± 11.4 | 0.021 ± 0.008 | 1893 ± 340 |
| EWF50_A 15 | 33.0 ± 2.9 | 0.019 ± 0.003 | 1935 ± 120 |
| EWF50_A 20 | 27.1 ± 6.0 | 0.021 ± 0.003 | 1690 ± 377 |
| EWF50_A 22 | 20.1 ± 5.9 | 0.020 ± 0.004 | 1313 ± 430 |
| EWF50_A 24 | 18.3 ± 3.2 | 0.018 ± 0.001 | 1121 ± 397 |
| EWF50_A 26 | 14.8 ± 2.2 | 0.016 ± 0.008 | 1188 ± 339 |
| EWF50_A 28 | 8.60 ± 1.4 | 0.012 ± 0.003 | 927 ± 136 |
| EWF50_A 30 | 5.84 ± 1.5 | 0.010 ± 0.002 | 705 ± 140 |

In addition, ANOVA was used to compare only the commercial system with the blend at 15 phr of ammine. With two levels, and one degree of freedom, for flexural strength, modulus, and strain at break, $p = 0.43$, $p = 0.8$, and $p = 0.61$ highlight that the two samples were not statistically different. This is consistent with the observation from DSC analysis, which showed that, for the sample at 15 phr amine, a relatively high Tg was found, which was only 15 °C lower than that of the commercial system. All the ANOVA findings are summarized in Tables 5–8.

**Table 5.** ANOVA parameters—amount of EWF as the source of variation.

| Property | Source of Variation | SS | MS | F-Test | p-Value |
|---|---|---|---|---|---|
| Flexural strength | EWF amount (4 levels) | 1244 | 414 | 8.48 | 0.0013 |
| | Error | 782 | 48.8 | | |
| Flexural modulus | EWF amount (4 levels) | $1.56 \times 10^6$ | $5.2 \times 10^5$ | 5.25 | 0.010 |
| | Error | $1.58 \times 10^6$ | $9.89 \times 10^4$ | | |
| Strain at break | EWF amount (4 levels) | 0.00017 | $5.7 \times 10^{-5}$ | 1.08 | 0.385 |
| | Error | 0.00084 | $5.3 \times 10^{-5}$ | | |

**Table 6.** ANOVA parameters—amount of EWF as the source of variation, neglecting the sample of commercial epoxy.

| Property | Source of Variation | SS | MS | F-Test | p-Value |
|---|---|---|---|---|---|
| Flexural strength | EWF amount (3 levels) | 6.53 | 3.26 | 0.149 | 0.862 |
| | Error | 262 | 21.85 | | |
| Flexural modulus | EWF amount (3 levels) | $1.36 \times 10^5$ | $6.82 \times 10^4$ | 0.73 | 0.502 |
| | Error | $1.12 \times 10^6$ | $9.34 \times 10^4$ | | |
| Strain at break | EWF amount (3 levels) | $3.3 \times 10^{-6}$ | $1.7 \times 10^{-6}$ | 0.0387 | 0.962 |
| | Error | 0.00052 | $4.3 \times 10^{-5}$ | | |

**Table 7.** ANOVA parameters—effect of the amount of amine on the mechanical properties.

| Property | Source of Variation | SS | MS | F-Test | p-Value |
|---|---|---|---|---|---|
| Flexural strength | Amine amount (8 levels) | 2791 | 465 | 33.1 | $1.84 \times 10^{-11}$ |
| | Error | 394 | 14 | | |
| Flexural modulus | Amine amount (8 levels) | $5.43 \times 10^6$ | $9.05 \times 10^5$ | 9.71 | $8.35 \times 10^{-6}$ |
| | Error | $2.61 \times 10^6$ | $9.31 \times 10^4$ | | |
| Strain at break | Amine amount (8 levels) | $5.24 \times 10^{-4}$ | $8.6 \times 10^{-5}$ | 5.40 | $8.18 \times 10^{-4}$ |
| | Error | $4.56 \times 10^{-4}$ | $1.6 \times 10^{-5}$ | | |

**Table 8.** ANOVA parameters—effect of the amount of amine on the mechanical properties, neglecting samples with commercial epoxy.

| Property | Source of Variation | SS | MS | F-Test | p-Value |
|---|---|---|---|---|---|
| Flexural strength | Amine amount (2 levels) | 48.4 | 48.4 | 0.699 | 0.427 |
| | Error | 553 | 69.2 | | |
| Flexural modulus | Amine amount (2 levels) | 4410 | 4410 | 0.0678 | 0.801 |
| | Error | $5.20 \times 10^5$ | $6.5 \times 10^4$ | | |
| Strain at break | Amine amount (2 levels) | 0.00001 | 0.00001 | 0.274 | 0.615 |
| | Error | 0.00029 | $3.7 \times 10^{-5}$ | | |

The effect of the lower Tg of the blends on the flexural strength and modulus is also highlighted in Figure 12a,b. From the plots, it is clear that all the samples produced at different EWF or amine amounts fall on a single master curve. Interestingly, also the commercial epoxy falls on the same master curve. This indicates that the low-range Tg is the parameter that mainly influenced the mechanical properties of the produced resin. The higher-range Tg (measured around 150 °C) and the chemical structure of the resulting polymer had only marginal effects on the flexural strength and modulus of the produced blends.

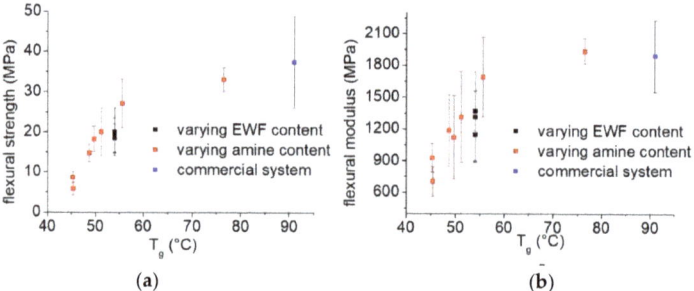

**Figure 12.** Effect of the lower-range Tg on the evolution of (**a**) flexural strength and (**b**) flexural modulus.

The recycled polymer obtained from the chemical recycling of the epoxy network was characterized by DMA and DSC. The tan δ vs. temperature was measured at three different frequencies: 1, 10, and 30 Hz. The results (Figure 13) clearly showed the presence of a single peak for all the frequencies tested. The peak was centered at 69 °C at 1 Hz, and it shifted to higher temperatures for increasing frequencies. This relaxation behavior was shown by glass transition temperature (Tg). Similar values were typically displayed by similar polymers derived from the recycling of bio-based epoxies cured by cleavable ammines [12]. Reprocessable bio-based epoxy cured using an aromatic disulfide crosslinker with diacid functionality displayed, after recycling, a glass transition temperature between 65 and 73 °C [26]. Slightly higher (i.e., 80 °C) and even lower (i.e., 18 °C) Tg values were obtained by the same group, varying the epoxy precursor among different epoxidized linseed and soybean oils cured by 2,2′-dithiodibenzoic acid [27].

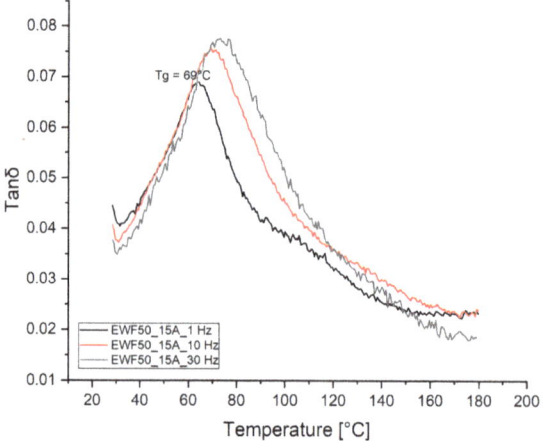

**Figure 13.** Tan δ vs. temperature of the polymer obtained from the recycling of the EWF50_A15 formulation.

## 4. Conclusions

An exemplary bio-based epoxy resin formulation showing full recyclability was presented in this paper. Two different epoxy resin precursors were mixed with a cleavable ammine: a commercial bio-based epoxy and a novel epoxidized waste flour. The latter chemical reactant was obtained using a green approach based on the use of a UV/ozone treatment. Mixing the two epoxy monomers allowed for obtaining formulations easy to mix at room temperature showing low viscosity in the unreacted state that could be cured at temperatures varying between 120, 150, and 160 °C.

The glass transition temperature of the epoxy formulation was optimized by varying the ratio between the two epoxy monomers and the amount of the cleavable amine. When the ammine was added at 22 phr, the glass transition temperature was fixed at about 54 °C with no significant change varying the bio-based epoxy content. However, reducing the ammine content down to 15 phr improved the glass transition temperature up to 76.5 °C, while increasing the ammine content to 30 phr reduced the Tg down to 45.4 °C. Similar results were obtained when considering the mechanical properties with the formulation cured with 15 phr ammine content. This hardener amount was chosen as optimal since a further reduction of the ammine content resulted in high viscosity systems.

The resin developed showed optimal recyclability when treated with an acidic solution under mild conditions (i.e., 80 °C for 1 h), obtaining a plastic material with a Tg of 69 °C by DMA and about 50 °C by DSC. The plastic material obtained from the recycling is unique compared with those obtained before as it was derived from a valuable thermoset formulated using 50 wt% of an epoxy monomer synthetized from waste flours.

The use of materials derived from organic waste as valuable thermoset prepolymer, which, after curing, can also be recycled at the end of their life, can truly revolutionize the recycling strategies for organic wastes, avoiding negative impacts on the environment. At the same time, the possibility to recover a high percentage (i.e., 85%) of this product into a reusable plastic matrix can guarantee the respect of the environment. However, more efforts will be required to further optimize the thermal and mechanical properties of the cured thermosets. These efforts should be focused on the synthesis of reactants, leading to a balanced stoichiometry and with a stiffer structure to improve the glass transition temperature and the mechanical properties.

## 5. Patents

Esposito Corcione, C., Greco, A., Visconti, P., Striani, R., Ferrari, F., 2019. Process for the production of bio-resins and bio-resins thus obtained. 102019000016151. IT.

**Author Contributions:** F.F., R.S. and L.S. performed the experiments; C.E.C., G.C. and A.G. supervised the work. All authors have read and agreed to the published version of the manuscript.

**Funding:** This research did not receive any specific grant from funding agencies in the public, commercial, or not-for-profit sectors.

**Institutional Review Board Statement:** Not applicable.

**Informed Consent Statement:** Not applicable.

**Data Availability Statement:** Not applicable.

**Acknowledgments:** Gianluca Cicala acknowledges the project Thalassa (PON "R&I00 2014e2020", grant no. ARS01_00293).

**Conflicts of Interest:** The authors declare no conflict of interest.

## References

1. Zotti, A.; Zuppolini, S.; Borriello, A.; Zarrelli, M. Thermal and mechanical characterization of an aeronautical graded epoxy resin loaded with hybrid nanoparticles. *Nanomaterials* **2020**, *10*, 1388. [CrossRef] [PubMed]
2. Awad, S.A.; Mahini, S.S.; Fellows, C.M. Modification of the resistance of two epoxy resins to accelerated weathering using calcium sulfate as a photostabilizer. *J. Macromol. Sci. Part A* **2019**, *56*, 316–326. [CrossRef]

3.  La Rosa, A.D.; Cicala, G. *Handbook of Life Cycle Assessment (LCA) of Textiles and Clothing*; Woodhead Publishing Series in Textiles; Muthu, S.S., Ed.; Woodhead Publishing: Sawston, UK, 2015; pp. 301–323. ISBN 9780081001691.
4.  Rybicka, J.; Tiwari, A.; Leeke, G.A. Technology readiness level assessment of composites recycling technologies. *J. Clean. Prod.* **2016**, *112*, 1001–1012. [CrossRef]
5.  Giorgini, L.; Benelli, T.; Brancolini, G.; Mazzocchetti, L. Recycling of carbon fiber reinforced composites waste to close their Life Cycle in a Cradle-to-Cradle approach. *Curr. Opin. Green Sustain. Chem.* **2020**, *26*, 100368. [CrossRef]
6.  Zhang, J.; Chevali, V.S.; Wang, H.; Wang, C.H. Current status of carbon fibre and carbon fibre composites recycling. *Compos. Part B Eng.* **2020**, *193*, 108053. [CrossRef]
7.  Pimenta, S.; Pinho, S.T. Recycling carbon fibre reinforced polymers for structural applications: Technology review and market outlook. *Waste Manag.* **2011**, *31*, 378–392. [CrossRef]
8.  Li, J.; Xu, P.-L.; Zhu, Y.-K.; Ding, J.-P.; Xue, L.-X.; Wang, Y.-Z. A promising strategy for chemical recycling of carbon fiber/thermoset composites: Self-accelerating decomposition in a mild oxidative system. *Green Chem.* **2012**, 3260–3263. [CrossRef]
9.  Wang, Y.; Cui, X.; Ge, H.; Yang, Y.; Wang, Y.; Zhang, C.; Li, J.; Deng, T.; Qin, Z.; Hou, X. Chemical Recycling of Carbon Fiber Reinforced Epoxy Resin Composites via Selective Cleavage of the Carbon–Nitrogen Bond. *ACS Sustain. Chem. Eng.* **2015**, *3*, 3332–3337. [CrossRef]
10. Cicala, G.; La Rosa, A.D.; Latteri, A.; Banatao, R.; Pastine, S. The use of recyclable epoxy and hybrid lay up for biocomposites: Technical and LCA evaluation. In Proceedings of the CAMX 2016—Composites and Advanced Materials Expo, Anaheim, CA, USA, 26–29 September 2016.
11. Cicala, G.; Mannino, S.; La Rosa, A.D.; Banatao, D.R.; Pastine, S.J.; Kosinski, S.T.; Scarpa, F. Hybrid biobased recyclable epoxy composites for mass production. *Polym. Compos.* **2017**. [CrossRef]
12. Cicala, G.; Pergolizzi, E.; Piscopo, F.; Carbone, D.; Recca, G. Hybrid composites manufactured by resin infusion with a fully recyclable bioepoxy resin. *Compos. Part B Eng.* **2018**, *132*, 69–76. [CrossRef]
13. Daniela, A.; Rosa, L.; Greco, S.; Tosto, C.; Cicala, G. LCA and LCC of a chemical recycling process of waste CF-thermoset composites for the production of novel CF-thermoplastic composites. Open loop and closed loop scenarios. *J. Clean. Prod.* **2021**, *304*, 127158.
14. Tan, S.G.; Chow, W.S. Biobased epoxidized vegetable oils and its greener epoxy blends: A review. *Polym.-Plast. Technol. Eng.* **2010**, *49*, 1581–1590. [CrossRef]
15. Ma, S.; Liu, X.; Jiang, Y.; Tang, Z.; Zhang, C.; Zhu, J. Bio-based epoxy resin from itaconic acid and its thermosets cured with anhydride and comonomers. *Green Chem.* **2013**, *15*, 245–254. [CrossRef]
16. Aouf, C.; Nouailhas, H.; Fache, M.; Caillol, S.; Boutevin, B.; Fulcrand, H. Multi-functionalization of gallic acid. Synthesis of a novel bio-based epoxy resin. *Eur. Polym. J.* **2013**, *49*, 1185–1195. [CrossRef]
17. Llevot, A.; Grau, E.; Carlotti, S.; Grelier, S.; Cramail, H. From Lignin-derived Aromatic Compounds to Novel Biobased Polymers. *Macromol. Rapid Commun.* **2016**, *37*, 9–28. [CrossRef] [PubMed]
18. Esposito Corcione, C.; Ferrari, F.; Striani, R.; Visconti, P.; Greco, A. Recycling of organic fraction of municipal solid waste as an innovative precursor for the production of bio-based epoxy monomers. *Waste Manag.* **2020**, *109*, 212–221. [CrossRef]
19. Ferrari, F.; Striani, R.; Minosi, S.; de Fazio, R.; Visconti, P.; Patrono, L.; Catarinucci, L.; Esposito Corcione, C.; Greco, A. An innovative IoT-oriented prototype platform for the management and valorization of the organic fraction of municipal solid waste. *J. Clean. Prod.* **2020**, *247*, 119618. [CrossRef]
20. Esposito Corcione, C.; Greco, A.; Visconti, P.; Striani, R.; Ferrari, F. Process for the Production of Bio-Resins and Bio-Resins thus Obtained. IT 102019000016151, 2019.
21. La Rosa, A.D.; Banatao, D.R.; Pastine, S.J.; Latteri, A.; Cicala, G. Recycling treatment of carbon fibre/epoxy composites: Materials recovery and characterization and environmental impacts through life cycle assessment. *Compos. Part B Eng.* **2016**, *104*, 17–25. [CrossRef]
22. Pastine, S.J. Sterically Hindered Aliphatic Polyamine Cross-Linking Agents, Compositions Containing and Uses Thereof. U.S. Patent 9,862,797, 9 January 2018.
23. Mahlin, D.; Wood, J.; Hawkins, N.; Mahey, J.; Royall, P.G. A novel powder sample holder for the determination of glass transition temperatures by DMA. *Int. J. Pharm.* **2009**, *371*, 120–125. [CrossRef] [PubMed]
24. Cicala, G.; Mamo, A.; Recca, G.; Restuccia, C.L. Synthesis and thermal characterization of some novel ABA block copolymers. *Macromol. Mater. Eng.* **2007**, *292*, 588–597. [CrossRef]
25. Carlier, V.; Sclavons, M.; Legras, R. Supported dynamic mechanical thermal analysis: An easy, powerful and very sensitive technique to assess thermal properties of polymer, coating and even nanocoating. *Polymer (Guildf.)* **2001**, *42*, 5327–5335. [CrossRef]
26. Di Mauro, C.; Genua, A.; Rymarczyk, M.; Dobbels, C.; Malburet, S.; Graillot, A.; Mija, A. Chemical and mechanical reprocessed resins and bio-composites based on five epoxidized vegetable oils thermosets reinforced with flax fibers or PLA woven. *Compos. Sci. Technol.* **2021**, *205*, 108678. [CrossRef]
27. Di Mauro, C.; Malburet, S.; Graillot, A.; Mija, A. Recyclable, Repairable, and Reshapable (3R) Thermoset Materials with Shape Memory Properties from Bio-Based Epoxidized Vegetable Oils. *ACS Appl. Bio Mater.* **2020**, *3*, 8094–8104. [CrossRef]

MDPI

St. Alban-Anlage 66

4052 Basel

Switzerland

Tel. +41 61 683 77 34

Fax +41 61 302 89 18

www.mdpi.com

*Polymers* Editorial Office

E-mail: polymers@mdpi.com

www.mdpi.com/journal/polymers